A Many-Sorted Calculus Based on Resolution and Paramodulation

Christoph Walther
Institut fur Informatik I
Karlsruhe University

A Many-Sorted Calculus Based on Resolution and Paramodulation

Pitman, London

Morgan Kaufmann Publishers, Inc., Los Altos, California

PITMAN PUBLISHING
128 Long Acre, London WC2E 9AN

First published 1987

Available in the Western Hemisphere from
MORGAN KAUFMANN PUBLISHERS, INC.,
95 First Street, Los Altos, California 94022

ISSN 0268-7526

British Library Cataloguing in Publication Data
Walther, Christoph
A many-sorted calculus based on
resolution and paramodulation.—(Research
notes in artificial intelligence,
ISSN 0268-7526).
1. Artificial intelligence 2. Reasoning
I. Title II. Series
006.3 Q335

ISBN 0-273-08718-5

Library of Congress Cataloging in Publication Data
Walther, Christoph.
A many-sorted calculus based on resolution and
paramodulation.

(Research notes in artificial intelligence)
Bibliography: p.
Includes index.
1. Automatic theorem proving. 2. Artificial
intelligence. I. Title. II. Series: Research notes
in artificial intelligence (London, England)
QA76.9.A96W35 1987 511.3 87-2815
ISBN 0-934613-49-4 (Morgan Kaufmann)

Reproduced and printed by photolithography
in Great Britain by Biddles Ltd, Guildford

Contents

Notation

Throughout this monograph the following notation is used:

id — identity function

$f_{|M}$ — function f restricted to a subset M of its domain

$f(t)\downarrow$ — t is in the domain of the function f

$f \in (M \to N)$ — f is a function from M to N

$\circ$ — composition of functions

$|$ — negation, e.g. $x \not\leq y$ means not $x \leq y$

$|M|$ — cardinality of set M

$M \times N$ — cartesian product of M and N

$M \setminus N$ — set theoretic difference of M and N

$M-L$ — abbreviates $M \setminus \{L\}$

⊘ — end of case in a proof by cases

⊠ — end of an example, definition or proof

∇ — contradiction

Preface

This monograph is a revised and translated version of a doctoral dissertation accepted by the Department of Computer Science at the University of Karlsruhe in July 1984. It is concerned with the utilization of the advantages and concepts of many-sorted logic for resolution and paramodulation based automated theorem proving. Most of the material presented here has been prepared during the years 1980-1983 when I was a member of the theorem proving research group developing the Markgraf Karl Refutation Procedure.

I would like to thank Prof. Dr. P. Deussen, who consented to become my supervisor, for his support. I am grateful to Jörg Siekmann, my second supervisor, who first aroused my interest in automated theorem proving and who provided an opportunity for me to participate in the Markgraf Karl Project in an intellectually stimulating atmosphere. His criticism, support and encouragement greatly contributed to this work.

I am also indebted to Werner Stephan, who formulated the idea of many-sorted automated theorem proving, when he become frustrated by Markgraf Karl's weak performance on some simple theorems from lattice theory, and to my colleagues from the early days of the Markgraf Karl Group, Karl-Hans Bläsius, Norbert Eisinger, Alexander Herold, and Hans Jürgen Ohlbach, for their support and many stimulating and useful discussions.

Finally, I would like to thank the editors and referees of the Research Notes for their helpful hints and suggestions, and the publisher for his assistance (and patience) in preparing this monograph.

"As a rule," said Holmes, "the more bizarre a thing is the less mysterious it proves to be. It is your commonplace, featureless cases which are really puzzling."

A. C. Doyle, The Red-Headed League

1 Introduction

A successful method humans use in reasoning is to assume a structured universe of discourse. We assume that the universe is divided into certain subuniverses, instead of having a single one. Certain relations exist between pairs of these subuniverses - they may be disjoint, they may have non-empty intersections or one may be completely contained in another. We also assume that certain objects belong to certain subuniverses and that certain mappings and relations are meaningful only for certain subuniverses.

In geometry, for instance, we may consider a universe of discourse, which is divided into **points**, **lines**, **angles**, **triangles**, **rectangles**, **polygons** etc., where we assume, for instance, that points and angles have no objects in common, or that triangles as well as rectangles are also polygons. Here we may think of specific points p_1, p_2, p_3, or of some specific angles α and β. We may also have a notion of **parallelism**, which of course is meaningful only for lines, but not for points, angles etc. We may associate an **area** with each polygon, where we assume that each area is a **real number**, i.e. another subuniverse, which is disjoint with each of the other ones. Of course, our implicit understanding of areas prevents us from talking of the area of a point or of an angle, and also from assuming that the area of a polygon can be a triangle.

Thinking of a structured universe helps humans to draw logical consequences efficiently. It also helps to avoid wasting our time with meaningless or wrong conclusions. Given, for instance, that the area of each polygon is a **positive** real number, we implicitly and immediately assume that the same holds true for triangles as well as for rectangles. But we will never conclude that a triangle is parallel to itself (which is meaningless) from the fact the each line is parallel to itself. We also do not assume that the sum of all angles in a polygon is 180° (which is wrong), despite the fact that this holds true for each triangle.

We think of structured universes in almost every branch of mathematics, and therefore it is not a mere accident that almost all mathematical textbooks are written in a language which reflects the specific (sub)universes of discourse, albeit often very implicitly.

This monograph is concerned with a mechanization of this aspect of human reasoning. Our starting point is some first-order calculus, which defines how we obtain theorems from given hypotheses by pure syntactic reasoning. This calculus is extended to a many-sorted calculus, i.e. a calculus which allows to make explicit all the implicit assumptions and notions discussed above. As we shall see, it is possible in this case to shift all the semantic and implicit argumentation to the syntactic and explicit level of formal first-order reasoning.

Many-Sorted Calculi The development of a many-sorted (*mehrsortig*) version of some given (sound and complete) first-order one-sorted or unsorted (*einsortig, unsortiert*) calculus is well known in the area of formal logic [Herbrand 1930, Schmidt 1938, Schmidt 1951, Wang 1952, Gilmore 1958, Hailperin 1957, Oberschelp 1962, Idelson 1964]. The first-order language has to be extended to a many-sorted language, the semantics for this language have to be defined, and the rules of inference have to be modified accordingly.

The Many-Sorted Language Our subuniverses of discourse are given names, called sort symbols, where the set of all sort symbols is partially ordered by the subsort order relation, thus expressing the inclusion relations which hold between the subuniverses under consideration. We may use, for instance, angle, triangle, polygon etc. as sort symbols, where we define triangle as a subsort of polygon. A set of sort symbols with the subsort order imposed on it is called a sort hierarchy.

Variable and function symbols (of our given calculus) are associated with a sort symbol, called the rangesort. The sort of a term is determined by the rangesort of its outermost symbol. This is a syntactic formulation of the fact that the (semantic) object represented by a given term is a member of the subuniverse represented by the sort of that term.

We also restrict the domains of function and predicate symbols to terms of certain sorts (one for each argument position), called the domainsorts for the respective argument positions. This syntactic requirement expresses the fact that not each term is meaningful as an argument to each function or predicate symbol. In geometry, for instance, we only want to have terms of sort line as arguments of a predicate parallel and we demand that only terms of sort polygon are used as arguments of the function area. The definitions of the range- and domainsorts for all the variable, function and predicate symbols under consideration are collected in a so-called (many-sorted) signature.

Consequently we demand that only those well-formed formulas of our given calculus are allowed in its many-sorted version, which are well-sorted (*sortenrecht*). In the construction of these well-sorted formulas we define that each variable and each constant is a well-sorted term and we allow for each argument position of a function or predicate symbol only well-sorted terms of the domainsort or of a subsort of this domainsort stipulated for the argument position of the respective function or predicate symbol. Hence, for instance, we can apply the function area to terms of sort polygon but also to terms of sort triangle or rectangle, provided both are subsorts of polygon. But we never use a term of sort angle as an argument of area.

Semantics Having defined the many-sorted language, i.e the set of all well-formed and well-sorted formulas, of our many-sorted calculus, we need a notion of truth for these formulas. Of course, the semantic notions have to mirror our assumptions on sorts as subuniverses and the subsort order as inclusion relation.

The necessary extensions of the semantic notions are straightforward: as intended, we demand that the domain of an interpretation is divided into certain subdomains. An interpretation associates each sort symbol with a certain subdomain, such that a subdomain is included in another (in the set-theoretical sense), if the sort symbol designating the included subdomain is a subsort of the sort symbol designating the other one.

Furthermore, an interpretation associates each function symbol with a mapping and each predicate symbol with a relation defined on the subdomains given by the interpretation for the respective domainsort symbols and yielding always an element of the subdomain associated with the rangesort in the case of a function symbol.

Finally, each variable assignment assigns a variable of a certain sort to an element of the subdomain associated with that sort. The remaining semantic notions, e.g. the semantics of the connectives and quantifiers, are the same as in the unsorted case. We let $\models_\Sigma \Phi$ denote the validity of a well-sorted formula and use HYP $\models_\Sigma \Phi$ to indicate that Φ is semantically implied by the set of hypotheses HYP, whose members are well-sorted formulas.

Inference Rules The inference rules of the many-sorted calculus correspond to the inference rules of the given calculus, but with the restriction that only well-sorted formulas can be deduced by an application of the restricted inference rules. Starting with well-sorted formulas this guarantees that only well-sorted formulas are derived in a deduction of the many-sorted calculus. We let $\vdash_\Sigma \Phi$ denote that Φ is a theorem of the many-sorted calculus and we write HYP $\vdash_\Sigma \Phi$ to indicate that there is a deduction of Φ from the set of hypotheses HYP in the many-sorted calculus.

Obviously we are only interested in a many-sorted calculus which is sound and complete, i.e. we have to choose our modifications of the given inference rules such that

(1) $\models_\Sigma \Phi$ iff $\vdash_\Sigma \Phi$, for each well-sorted formula Φ,

is guaranteed. Having defined a many-sorted calculus which satisfies (1), we have succeeded in shifting the implicit and semantic notions associated with our thinking of a structured universe of discourse, to a purely syntactic level of formal reasoning.

Efficiency of many-sorted reasoning But what have we gained and what are the advantages of using this new calculus ? The advantages can be seen if we compare the deductive capabilities of both calculi, i.e. if we use the given unsorted calculus instead of its many-sorted version to prove theorems.

To facilitate a comparison between both calculi, we need a means to express the sort of a term in the unsorted calculus: we use sort symbols as unary predicate symbols. Then we can explicitly represent the relations between the function symbols and the sort symbols as well as the subsort order by a set A^{Σ} of so called sort axioms (*Sortenaxiome*), i.e. a set of first-order (unsorted) formulas. For instance,

$$\forall x \text{ polygon}(x) \Rightarrow \text{real\_number}(\text{area}(x))$$

is a sort axiom expressing that area maps polygons into real numbers, and

$$\forall x \text{ triangle}(x) \Rightarrow \text{polygon}(x)$$

is the sort axiom for our assumption that triangles are polygons.

For a well-sorted formula Φ like, for instance,

$$\forall x{:}\text{line } \exists y{:}\text{line parallel}(xy)$$

the relativization $\hat{\Phi}$ (*Sortenbeschränkung, Relativierung*) Φ is the unsorted version of Φ, as e.g.,

$$\forall x \text{ line}(x) \Rightarrow (\exists y \text{ line}(y) \wedge \text{parallel}(xy))$$

where sort symbols are used as unary predicate symbols to express the sorts of the variables. Given the notions of sort axioms and relativizations, the relation between a first-order calculus and its many-sorted version can be described by the following diagram:

$$
\begin{array}{ccc}
 & (1) & \\
\models_{\Sigma}\Phi & \longleftrightarrow & \vdash_{\Sigma}\Phi \\
(2.1)\ \Big\updownarrow & & \Big\updownarrow\ (2.2) \\
A^{\Sigma}\models\hat{\Phi} & \longleftrightarrow & A^{\Sigma}\vdash\hat{\Phi} \\
 & (3) &
\end{array}
$$

Figure 1.1 Unsorted and many-sorted logic

The equivalences (1) and (3) are given by the soundness and completeness of both calculi, where $\models$ and $\vdash$ denote semantic implication and formal deduction in the unsorted calculus. The equivalences (2.1) and (2.2) form the so-called Sort-Theorem. Equivalence (2.1) is the modeltheoretic part and (2.2) is the prooftheoretic part of the Sort-Theorem.

Now we can see what advantages we have in using a many-sorted calculus: the prooftheoretic part (2.2) of the Sort-Theorem shows that a well-sorted formula Φ is a theorem of the many-sorted calculus iff its relativization $\hat{\Phi}$, which is more complex than Φ, can be deduced in the unsorted calculus from the set of sort axioms A^{Σ} as additional hypotheses. Hence we obtain a shorter deduction with smaller formulas from a smaller set of hypotheses, when proving $\vdash_{\Sigma}\Phi$ instead of $A^{\Sigma}\vdash\hat{\Phi}$. The reason is that deductions about sorts, which are performed explicitly in the unsorted calculus (using sort axioms and relativizations), are built into the inference mechanism in the many-sorted calculus.

But we can also see that in an unsorted calculus a lot of useless formulas can be deduced, which do not have a many-sorted counterpart, because these many-sorted formulas would be meaningless or wrong. Given, for instance, the axiom

$$\forall x{:}\text{line parallel}(xx)$$

and a term α of sort angle, we cannot deduce

$$\text{parallel}(\alpha\alpha)$$

in the many-sorted calculus, because this formula is not well-sorted and is therefore meaningless. But in the unsorted calculus we may obtain from the relativization

$$\forall x \ \text{line}(x) \Rightarrow \text{parallel}(xx)$$

by universal instantiation the formula

$$\text{line}(\alpha) \Rightarrow \text{parallel}(\alpha\alpha)$$

which is true but useless, since line(α) does not hold.

Also it is impossible to deduce from the given axiom

$$\forall x{:}\text{triangle angle\_sum}(x) \equiv 180^\circ$$

for any polygon p the well-sorted but wrong formula

$$\text{angle\_sum}(p) \equiv 180^\circ$$

whereas in the unsorted calculus we do not have any difficulties inferring the useless truth

$$\text{triangle}(p) \Rightarrow \text{angle\_sum}(p) \equiv 180^\circ$$

from the relativized axiom

$$\forall x \ triangle(x) \Rightarrow angle\_sum(x) \equiv 180^\circ.$$

These observations lead to the conclusion that a many-sorted calculus is a good foundation for automated theorem proving, because it is a syntactical device to draw logical consequences efficiently and to avoid wasting time with the derivation of useless conclusions.

° ° ° ° °

The basic ideas of many-sorted resolution and paramodulation based theorem proving are easy to understand and an implementation is obtained without severe difficulties. Certain deductions are decidable by the many-sorted unfication procedure. Without sorts, a theorem prover has to search for these deductions instead, giving rise to overhead and deadends.

An informal introduction of a resolution and paramodulation based many-sorted calculus is given in Chapter 2. This calculus is a generalization of the unsorted system, i.e. our many-sorted calculus is identical to its unsorted counterpart, if the set of sort symbols is a singleton (yielding a one-sorted calculus). The advantages of such a calculus and the problems related to its definition are illustrated with several examples. The formal notions for resolution and paramodulation based theorem proving are recalled in Chapter 3 and the notions for extending this calculus to a many-sorted one are presented in Chapter 4. The practical application of this calculus in automated theorem proving leads to a drastic reduction of the search space and to shorter refutations of smaller sets of shorter clauses. Chapter 11 describes all the modifications necessary to extend a resolution and paramodulation based system, yielding a many-sorted automated theorem prover. It can be seen that the advantages of this approach cause almost no additional costs. Chapter 12 presents some experimental results obtained with such a system. We conclude in Chapter 13 with a brief account of the historical development of many-sorted logic and also discuss some related work.

The remaining chapters constitute the theoretical part of this monograph (which, however, is not necessary to the understanding of the former one): many-sorted ground term rewriting is considered in Chapter 5, because ground term rewriting provides a formal tool to deal conveniently with E-interpretations (i.e. interpretations with "inbuilt" semantics for equality). Based on the results on many-sorted ground term rewriting, we can prove the ground completeness of our calculus in Chapter 6. The results on ground term rewriting and ground completeness are necessary, because many-sorted paramodulation can be considered as a real refinement of paramodulation, which is effective even in the ground case. In the general case, the presence of sorts influences the notion of unification. In Chapter 7 many-sorted unification is investigated and a many-sorted version of the Unification Theorem is proved. Using Ground Completeness and the results on many-sorted unification, we prove the completeness of the many-sorted calculus by the usual lifting argument in Chapter 8. The soundness of our calculus is shown in Chapter 9. We conclude the theoretical part in Chapter 10 with a proof of the Sort-Theorem, which connects the semantic and deductive notions of the many-sorted calculus with its unsorted counterpart.

2 Many-Sorted Resolution and Paramodulation

Given a first-order calculus and its many-sorted version, the Sort Theorem tells us that every truth which can be deduced in the many-sorted calculus, can also be deduced in the unsorted system (cf. Figure 1.1). But conversely, there are theorems of the unsorted calculus which do not have many-sorted counterparts. However, this does not entail an incompleteness property of the many-sorted system, because these many-sorted non-theorems are meaningless (regarding the semantics of many-sorted logic). Hence both calculi are equivalent regarding soundness and completeness, and this is the reason why many-sorted logic has received scarce attention in the field of formal logic.

Since logicans use a formal calculus to study the foundations, the nature, and the limitations of formal reasoning, there is a good reason to choose a calculus as simple as possible. Investigations are much simplified if all features which do not represent a fundamental idea (like, for instance, variables and quantifiers constitute the fundamental difference between propositional and first-order logic) are excluded from the formal system.

The situtation changes drastically, however, if we are concerned with automated theorem proving. Here a formal calculus is a device and a practical tool to find and compute proofs and theorems. Now efficiency is a concern (which it is not in formal logic) and therefore the kind of a calculus is of considerable interest, because the performance of a theorem proving system depends directly on the performance of the calculus it implements.

The resolution calculus of Robinson [Robinson 1965, Chang and Lee 1973, Loveland 1978] is a first-order calculus which has been proved to be useful and successful in automated theorem proving. This calculus was augmented with the paramodulation rule by Wos and Robinson [Wos and Robinson 1973], i.e. an inference rule designed to implement equality reasoning efficiently.

The outstanding features of the resolution calculus with paramodulation (RP-calculus for short) are its deduction incompleteness and the principle of most generality, which are the reasons for the efficiency of this calculus.

The RP-calculus is deduction incomplete, because not all valid formulas, i.e. valid clauses, can be deduced. As a consequence a theorem prover based on this calculus does not deduce all truths from a set of formulas, which hold independently from this set. This feature counts for efficiency, because it helps to keep the search space small.

But it is also impossible to deduce all (non-valid) formulas from a given set of formulas, which are semantically implied by this set. Again, this feature helps to keep the search space small and therefore helps to obtain an efficient theorem proving system.

However for each non-valid formula Φ, which is semantically implied by a set of formulas, a formula Φ' can be deduced in the RP-calculus, such that Φ' semantically implies Φ. This feature, known as principle of most generality (because Φ' is more general than Φ), entails the refutation completeness of the RP-calculus.

Our intention here is to extend the RP-calculus to a many-sorted calculus, such that all the useful features of the given system, viz. soundness, deduction incompleteness, the principle of most generality, and of course the refutation completeness, are maintained.

Many-Sorted Resolution and Factorization Having defined a many-sorted language as in Chapter 1, the definition of a many-sorted clause language is straightforward: a well-sorted atom is a well-sorted atomic formula, a well-sorted literal is a well-sorted atom or its negation and a well-sorted clause is a finite set of well-sorted literals.

The many-sorted versions of the resolution and factorization rule have to be defined such that only well-sorted clauses can be deduced from well-sorted clauses by the many-sorted inference rules. Since the inference rules of the RP-calculus are based on the key concept of unification, we have to focus our attention on the many-sorted version of this notion.

For the many-sorted RP-calculus (or ΣRP-calculus for short), we demand for all unifiers that (1) variables are only replaced by well-sorted terms and that (2) the sort of a term is always equal to or a subsort of the sort of the variable it replaces. Those unifiers are called well-sorted and it is easily seen that an application of a well-sorted unifier to a well-sorted term (well-sorted literal or well-sorted clause) always yields a well-sorted term (well-sorted literal or well-sorted clause).

Now many-sorted resolution and factorization is defined as in the unsorted case, but with the proviso that only well-sorted clauses and well-sorted most general unifiers are used when resolvents and factors of the ΣRP-calculus are computed. Let us illustrate the ΣRP-calculus and also its advantages by a simple example (extracted from Schubert's Steamroller problem which is discussed in Chapter 12) and consider the following statements:

Example 2.1

(i) Birds and snails are animals and (ii) there are some of each of them. (iii) Also there are some plants. (iv) Every animal likes to eat all the animals which do not like to eat all plants. (v) Birds do not like to eat snails. (vi) Therefore there is an animal that likes to eat all plants.

The problem is to show that statement (vi) is true, whenever statements (i)-(v) are assumed to be true. We start with a many-sorted axiomatization of the problem:

Our sort hierarchy consists of sorts B,S,A, and P denoting birds, snails, animals, and plants, respectively. We define B and also S as subsorts of A, thus obtaining an axiomatization of statement (i). We further assume, that there are constants b of sort B, s of sort S and p of sort P and so an axiomatization of statements (ii) and (iii) is also obtained. Although these constants are not used in a proof, we do need them for theoretical reasons. By the presence of these constants, our sorts cannot be 'empty', i.e. the subdomains associated with the sorts by any interpretation cannot be empty, thus guaranteeing the soundness of our derivations (cf. Chapter 9). Figure 2.1 is a diagram of the sort hierarchy under consideration.

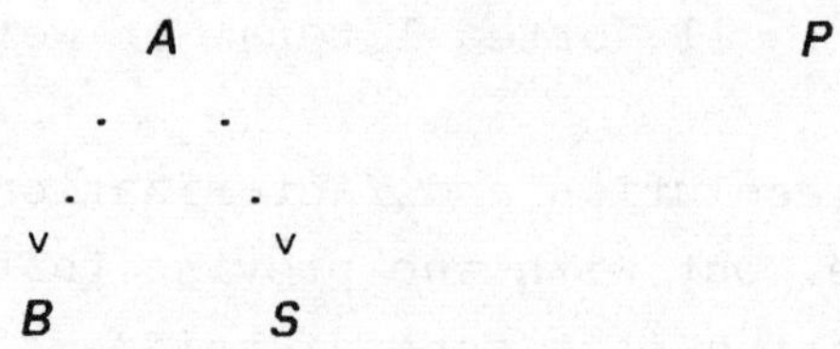

Figure 2.1 The sort hierarchy of Example 2.1

Using L(xy) as an abbreviation for 'x likes to eat y', the remaining statements are axiomatized by the well-sorted clauses shown in Figure 2.2.:

(1) $\{L(a_1 a_2), L(a_2 p_1)\}$ (2) $\{\neg L(b_1 s_1)\}$

(3) $\{\neg L(a_3 h(a_3))\}$

Figure 2.2 A many-sorted axiomatization for Example 2.1

Here h is a skolem function with domainsort A and rangesort P (mapping an animal to a plant it refuses to eat), which is necessitated by the existential quantification in the negated conclusion (vi). The subscripted lower case letters, e.g. a_1, a_2, p_1,... are universally quantified variables of the sorts denoted by the corresponding upper case letter, e.g. A,P,... .

We can resolve upon the literals 1(1) and 2(1) in Figure 2.2 using the well-sorted most general unifier $\{a_1 \leftarrow b_1, a_2 \leftarrow s_1\}$ (but not $\{b_1 \leftarrow a_1, s_1 \leftarrow a_2\}$, because this most general unifier violates the above condition (2) of well-sortedness). However, there is no such resolvent upon the literals 1(2) and 2(1), since there is no subsort relation between S and P and as a consequence the variables p_1 and s_1 of sort P and S are not unifiable under sorts.

From the well-sorted clause set of Figure 2.2, we obtain a solution of the problem by the following many-sorted refutation:

(R1) $\{L(s_1 p_1)\}$; many-sorted resolvent of 1(1) and 2(1)

(R2) $\square$; many-sorted resolvent of R1(1) and 3(1)

Here we use the well-sorted most general unifier $\{a_1 \leftarrow b_1, a_2 \leftarrow s_1\}$ for the computation of R1 and the well-sorted most general unifier $\{a_3 \leftarrow s_1, p_1 \leftarrow h(s_1)\}$ to obtain R2.

Now let us consider an unsorted formulation of the problem: Using the predicates

B(x) - x is a bird, S(x) - x is a snail,

A(x) - x is an animal, P(x) - x is a plant

as abbreviations, we obtain the set of clauses shown in Figure 2.3 as an unsorted clausal axiomatization of the problem.

(1) $\{\neg A(x), \neg A(y), \neg P(z), L(xy), L(yz)\}$ (2) $\{\neg B(x), \neg S(y), \neg L(xy)\}$

(3) $\{\neg A(x), \neg L(xh(x))\}$

(4) $\{\neg B(x), A(x)\}$ (5) $\{\neg S(x), A(x)\}$

(6) $\{B(b)\}$ (7) $\{S(s)\}$

(8) $\{P(p)\}$ (9) $\{\neg A(x), P(h(x))\}$

Figure 2.3 An unsorted axiomatization for Example 2.1

Here b,s and p are skolem constants necessitated by the existential quantifications in statements (ii) and (iii), and h is the skolem function as in Figure 2.2. Clauses (3) and (9) stem from the negation of statement (vi).

If we compare both axiomatizations, the advantages of the many-sorted approach become apparent: we only need 2 resolution steps for a refutation, compared to 7 steps in the unsorted case. Also the search space is reduced very much in the many-sorted case, because we have only 3 clauses with 4 literals, compared to 9 clauses with 19 literals. The reason is that sort information has to be represented explicitly in the unsorted case by **sort literals**, i.e. literals whose predicate letters are sort symbols. Using sort literals, we replace the given many-sorted clauses by their **relativizations** and extend the resulting clause set by the clausal representations for the sort axioms to obtain an unsorted clausal representation of the problem.

Note that the clauses (4) and (5) in Figure 2.3 are the sort axioms for the subsort order relation, clauses (6), (7) and (8) are the sort axioms stating the sorts of the constants and clause (9) is the sort axiom describing the domain- and the rangesort of the skolem function h. Clauses (1) - (3) in Figure 2.3 are the relativizations of the corresponding clauses in Figure 2.2, which are obtained by extending a well-sorted clause with all the sort literals of form ¬Q(x), where x is a variable of sort Q in the given well-sorted clause.

The resulting search space is further reduced by the constraints imposed on many-sorted unification. For instance, we can compute the resolvent R={¬A(x),¬A(y),¬P(z),¬B(y),¬S(z), L(xy)} upon the literals 1(5) and 2(3) in Figure 2.3 giving rise to the computation of several pure clauses: ¬S(z) can only be resolved upon 7(1), yielding the pure clause {¬A(x),¬A(y), ¬P(s),¬B(y),L(xy)} (because ¬P(s) cannot be resolved upon). Also ¬P(z) in R can only resolved upon the literals 8(1) and 9(2) yielding the pure literals ¬S(p) and ¬S(h(w)) respectively.

In the many-sorted case these deadends are impossible: the resolution step upon the literals 1(2) and 2(1) in Figure 2.2, which corresponds to the computation of R, is blocked, because the variables s_1 and p_1 have no well-sorted unifier.

As a result the size of the initial search space is reduced to 2 many-sorted resolvents and no many-sorted factors, compared to an initial search space of 18 resolvents and 2 factors in the unsorted case.

This example reveals the advantages of the ΣRP-calculus compared to the unsorted RP-calculus: we obtain a shorter refutation of shorter clauses from a smaller set of clauses and also a reduction of the search space.

The ΣRP-calculus is sound, because only well-sorted most general unifiers are used. But unfortunately this calculus, as defined so far, is (refutation) complete only if the sort hierarchy under consideration forms a forest structure, i.e. a finite collection of finite trees (like the sort hierarchy in Figure 2.1).

The ΣRP-calculus is (refutation) incomplete if non-forest structured sort hierarchies are used, because the Unification Theorem does not hold for this case. We shall illustrate this with another example:

Example 2.2

(i) Reptiles and birds are animals, which breathe by lungs and lay eggs.(ii) There are reptiles as well as birds. (iii) Each animal with lungs and wings has feathers. (iv) All animals without wings are poikilothermic, provided they lay eggs. (v) There is an animal which has feathers and lays eggs, or there is a poikilotherm, which breathes by lungs.

We give a many-sorted axiomatization expressing that statement (v) is true, whenever statements (i) - (iv) are assumed to be true: our sort hierarchy consists of sorts R,B,L and E denoting reptiles, birds, animals with lungs and animals which lay eggs, respectively. We define R and B as subsorts of L and also of E, thus obtaining a sortal representation of statement (i). We can do so by presence of statement (ii), which guarantees that neither R nor B are empty sorts (and by the subsort order L and E consequently are also non-empty). Figure 2.4 is a diagram of this sort hierarchy, which obviously is not a forest structure.

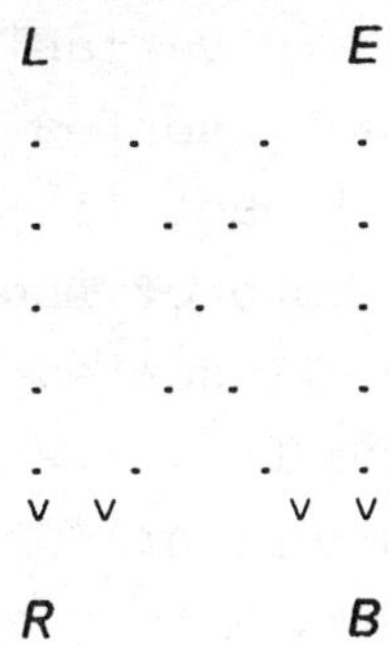

Figure 2.4 The sort hierarchy of Example 2.2

Using the predicates

$W(x)$ - animal x has wings, $F(x)$ - animal x has feathers,
$P(x)$ - animal x is poikilothermic

as abbreviations, we obtain the following many-sorted clausal representation of the problem:

(1) $\{\neg W(l_1), F(l_1)\}$ (2) $\{W(e_1), P(e_1)\}$

(3) $\{\neg F(e_2)\}$ (4) $\{\neg P(l_2)\}$

Figure 2.5 A many-sorted axiomatization for Example 2.2

Clauses (1) and (2) are the formal representations of statements (iii) and (iv), and clauses (3) and (4) stem from the negation of statement (v). As in the previous example, the subscripted lower case letters l_i and e_i (i=1,2) are universally quantified variables of the sorts denoted by the corresponding upper case letter L and E.

Now, assume that r_1 and b_1 are variables of sort R and B respectively. Then $\{l_1,e_1\}$ has well-sorted unifiers, e.g. $\{l_1 \leftarrow r_1, e_1 \leftarrow r_1\}$ and also $\{l_1 \leftarrow b_1, e_1 \leftarrow b_1\}$. But $\{l_1,e_1\}$ has no well-sorted most general unifier, because both of the most general unifiers $\{l_1 \leftarrow e_1\}$ and $\{e_1 \leftarrow l_1\}$ are not well-sorted. As a consequence, the ΣRP-calculus - as defined so far - is refutation incomplete, because we cannot compute a many-sorted resolvent from the set of clauses in Figure 2.4, although this set is unsatisfiable.

We shall overcome this problem with a technique borrowed from unification under equational theories: we use complete and minimal sets of well-sorted unifiers instead of most general unifiers, where the completeness of these sets entails the refutation completeness of the calculus and the minimality of these sets guarantees that the modified many-sorted calculus still satisfies the principle of most generality. In the above example, for instance, $\{\{l_1 \leftarrow r_1, e_1 \leftarrow r_1\},\{l_1 \leftarrow b_1, e_1 \leftarrow b_1\}\}$ is a minimal and complete set of well-sorted unifiers for $\{l_1,e_1\}$.

With this extension, we obtain the following many-sorted refutation from the clause set in Figure 2.5:

(R1) $\{\neg W(r_1)\}$; many-sorted resolvent of 1(2) and 3(1)

(R2) $\{P(r_1)\}$; many-sorted resolvent of 2(1) and R1(1)

(R3) $\square$; many-sorted resolvent of 4(1) and R2(1)

We use the first element of the minimal and complete set of well-sorted unifiers $\{\{l_1 \leftarrow r_1, e_2 \leftarrow r_1\},\{l_1 \leftarrow b_1, e_2 \leftarrow b_1\}\}$ for the computation of R1, and the sets $\{\{e_1 \leftarrow r_1\}\}$ and $\{\{l_2 \leftarrow r_1\}\}$ for the computation of R2 and R3, respectively.

However, since a complete and minimal set of unifiers can be arbitrarily large (although always finite, if the set of sort symbols is finite) it is possible to have an arbitrarily large number of many-sorted resolvents (or factors) for a given pair of clauses and literals resolved upon (or a given subset of a clause). Assume, for instance, that sorts s_1 and s_2 have 1,000,000 common but incomparable subsorts and that the variables x and y are of sort s_1 and sort s_2 respectively. Then we find that each complete and minimal set of well-sorted unifiers of {x,y} has 1,000,000 members, and as a consequence a clause like {P(x),P(y)} has 1,000,000 independent many-sorted factors.

If we ignore for the moment that this example is rather artificial, we may consider this effect as a possible disadvantage of the ΣRP-calculus, because the search space of an automated theorem prover can be swamped with a huge number of clauses by a single inference step. But if we use the unsorted calculus instead, we have 2,000,000 sort axioms in this example, each of which has to be used to compute the 1,000,000 relativizations of all the many-sorted factors. Hence the size of the search space would be even larger in the RP-calculus.

But there is a simple technical remedy to this kind of problem: if we embed the given sort hierarchy into a **meet-semilattice** by invention of additional sorts, we find that each complete and minimal set of well-sorted unifiers is always a **singleton**. For the sort hierarchy of Figure 2.4, for instance, we obtain the meet-semilattice by invention of a sort X, representing the union of reptiles and birds, and a sort Ø. Figure 2.6 is a diagram of the extended sort hierarchy.

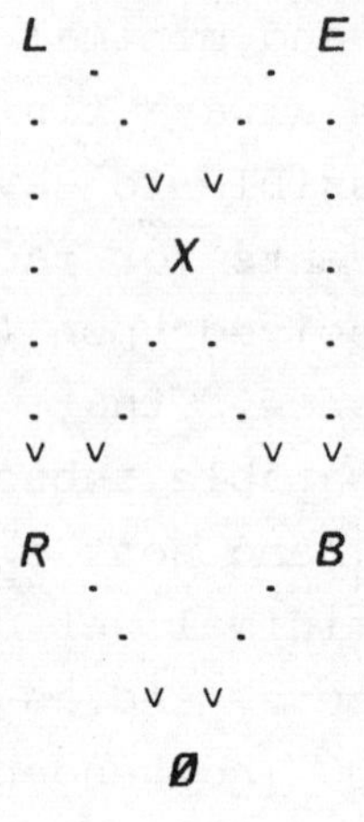

Figure 2.6 A meet semi-lattice for the sort hierarchy of Example 2.2

Now $\{l_1, e_1\}$ has the singleton $\{\{l_1 \leftarrow x_1, e_1 \leftarrow x_1\}\}$ as a complete and minimal set of well-sorted unifiers. The sort Ø is needed here only for formal reasons, i.e. to obtain a meet-semilattice, if there are sorts in the given hierarchy without common subsorts, like R and B. Since we assume that there are no terms of sort Ø, terms without common subsorts never have a well-sorted unifier.

But we prefer partially ordered sets of sort symbols instead of meet-semilattices in our definition of the ΣRP-calculus, because many-sorted axiomatizations tend to be unnatural or unintuitive (like the union X of reptiles and birds), if only meet-semilattices can be used to define a sort hierarchy.

Many-Sorted Paramodulation Paramodulation is an inference rule which integrates equality reasoning into the resolution calculus. But before we define a many-sorted version of this inference rule, we have to stipulate the domainsorts for the equality predicate ≡ : when concerned with equality reasoning, we assume that each sort hierarchy has a least upper bound, say any, and we define that any is the domainsort for both argument positions of ≡. Hence for all well-sorted terms q and r, q≡r is a well-sorted atom.

But this definition seems to ignore the information given by the sorts of the terms, because terms of arbitrary sort can be in the equality relation. One feels that it is more natural for a many-sorted logic to have the sorts of equal terms related somehow by the subsort order relation.

Given the equality q≡r, it seems meaningful to have that the sort of q is a subsort of, or equal to, the sort of r (or vice versa). A more liberal restriction would be to demand that the sorts of q and r have at least one common subsort, because the subdomains associated with the sorts of the terms must have a non-empty intersection, unless the equality is unsatisfiable.

However, those sortal restrictions for the equality predicate entail (implicitly) a strong assumption on the semantics of many-sorted logic in general and the sort hierarchies under consideration in particular. We suppose in this case, that if sort s_1 is not a subsort of sort s_2, the subdomain associated with s_1 by any interpretation is never included in the subdomain associated with s_2, and that the subdomains associated with s_1 and s_2 always have an empty intersection if s_1 and s_2 do not have a common subsort. Also the interpretation of a term t cannot be an element of the subdomain associated with a sort symbol s, provided the sort of t is not a subsort of s. Hence, as opposed to ordinary first-order logic, we assume certain propositions to be false by **default**, unless the contrary is stated explicitly.

We dispense with sortal restrictions on the domainsorts of the equality predicate, because many-sorted logic with default assumptions would not be compatible with unsorted logic. Also, this would require that sort hierarchies and signatures are complete in the sense that they mirror all inclusion and membership relations on the syntactic level, which may hold on the semantic level.

As opposed to many-sorted resolution and factorization, we do not neccesarily obtain a well-sorted clause if we only paramodulate with well-sorted clauses using well-sorted unifiers. Whenever the sorts of the terms in an equality disagree, it may happen that a term in a literal is replaced by a term whose sort is neither a subsort of, nor equal to, the domainsort of the specific argument position it occupies. Of course, we cannot allow such deductions of ill-sorted clauses in the ΣRP-calculus. Hence, in the case of paramodulation (and in contrast to resolution and factorization), the derivations of the many-sorted system are influenced by the domainsorts stipulated for the function and predicate symbols.

A straightforward idea to overcome this problem is to define a many-sorted paramodulant as a paramodulant generated from well-sorted clauses (using a member of a minimal and complete set of well-sorted unifiers), which in addition is a well-sorted clause.

Since the empty substitution is the only unifier in the ground case (and, of course, this unifier is well-sorted), this restriction is effective already in the ground case. But for the ground case completeness is maintained, because we can always assemble the 'right' equations by paramodulating into equalities (cf. Example 8.2). Unfortunately this method does not work for the general case, as can be seen from the following example:

Example 2.3

(i) Reptiles and birds are animals, which breathe by lungs and lay eggs. (ii) There are reptiles as well as birds. (iii) There is a bird which hatches out the eggs it lays. (iv) There is a reptile or a bird which does not hatch out its eggs. (v) There are at least two animals which breathe by lungs.

The problem is to show that statement (v) is true, whenever statements (i)-(iv) are assumed to be true. We use H(x) as an abbreviation for "animal x hatches out its eggs", where we define sort E as the domainsort for the unary predicate H, because "hatch out eggs" is a meaningful characterization only for animals which lay eggs. With the same sort hierarchy as given for Example 2.2 (cf. Figure 2.4), we obtain the following many-sorted clausal representation of the problem:

(1) $\{H(b')\}$ (2) $\{\neg H(b), \neg H(r)\}$

(3) $\{l_1 \equiv l_2)\}$

Figure 2.7 A many-sorted axiomatization for Example 2.3

Clauses (1) and (2) are the formal representations of statements (iii) and (iv), and clause (3) stems from the negation of statement (v). The symbols b',b and r are skolem constants of sort B and R respectively, which are necessitated by the existental quantifications in statements (iii) and (iv). As in the previous example, l_1 and l_2 are universally quantified variables of the sort L.

From the clause set in Figure 2.7, 6 paramodulants can be computed in the first generation, viz. $\{H(l_i)\}$, $\{\neg H(l_i), \neg H(r)\}$ and $\{\neg H(b), \neg H(l_i)\}$ (where i=1,2). But each of these paramodulants is ill-sorted, because L (i.e. the sort of l_i) is neither a subsort nor equals E (i.e. the domainsort of H) and therefore

$H(l_i)$ is not a well-sorted atom. Since there are no many-sorted resolvents and factors, a many-sorted refutation does not exist for the above clause set.

As can be seen from this example, the ΣRP-calculus would be (refutation) incomplete if a many-sorted paramodulant were defined as a paramodulant of the RP-calculus which is a well-sorted clause. But which necessary and meaningful conclusions should be drawn from the clauses in Figure 2.7, which cannot be obtained with the given restriction ?

We infer from our sort hierarchy and from the semantics of many-sorted logic, that everything which is true for animals with lungs, must hold for birds and also for reptiles. Hence we know that the clauses

(3.1) $\{l_1 \equiv b_1\}$ (3.2) $\{l_1 \equiv r_1\}$

must be true, provided clause (3) is true (where b_1 and r_1 are variables of sort B and R respectively). Replacing clause (3) by the clauses (3.1) and (3.2) in the above set, we obtain the following many-sorted refutation from the modified clause set:

(P1) $\{H(b_1)\}$; many-sorted paramodulant of 1(1) and 3.1(1)

(P2) $\{H(r_1)\}$; many-sorted paramodulant of 1(1) and 3.2(1)

(R3) $\{\neg H(r)\}$; many-sorted resolvent of P1(1) and 2(1)

(R4) □ ; many-sorted resolvent of P2(1) and R3(1)

To compute this deduction, we use the well-sorted unifiers $\{l_1 \leftarrow b'\}$, $\{l_1 \leftarrow b'\}$, $\{b_1 \leftarrow b\}$ and $\{r_1 \leftarrow r\}$, each of which is the only member of the respective complete and minimal set of unifiers.

Formally, we obtain clauses (3.1) and (3.2) by replacing the variable l_2 in clause (3) by the variables b_1 and r_1 respectively. Such a variable replacement is sound, i.e. yields a well-sorted clause, provided a variable is replaced by a variable the sort of which is a subsort of the sort of the replaced variable. We call the application of a sound variable replacement coercion, similar to the coercion operations in programming languages with types. If we allow coercions as additional inference rules, we are now able to infer all well-sorted clauses, which are necessary for a many-sorted refutation with paramodulation.

However, an unrestricted usage of coercions would destroy the principle of most generality. Assume, for instance, that we extend our given sort hierarchy by sorts S and D, standing for snakes and ducks, such that S is a subsort of R and D is a subsort of B. Consequently, S and D are also subsorts of L, and we obtain by unrestricted coercion

(3.3) $\{l_1 \equiv s_1\}$ (3.4) $\{l_1 \equiv d_1\}$

(where s_1 and d_1 are variables of sort S and D respecively). But obviously, we do not need clauses (3.3) and (3.4) for a many-sorted refutation of the above problem, because we can use clauses (3.1) and (3.2). Hence we infer more well-sorted instances of clause (3) than we need to obtain a solution.

Moreover, we want to have coercions only when they are necessary for a refutation at all, i.e. to enable the computation of a well-sorted paramodulant, which is necessary for a solution. Assume, for instance, we want to prove, that "there are at least two animals, which lay eggs" from the statements (i)-(iv) of Example 2.3. Then we obtain

(3') $\{e_1 \equiv e_2\}$

instead of clause (3), and each of the 6 well-sorted paramodulants $\{H(e_i)\}$, $\{\neg H(e_i),\neg H(r)\}$ and $\{\neg H(b),\neg H(e_i)\}$ (i=1,2) leads to a many-sorted refutation of the clauses (1),(2) and (3') without any usage of coercions.

Technically, we map the problem of finding most general and necessary coercions into a many-sorted unification problem: whenever a term is replaced by a term r in a paramodulation step, we create a fresh variable z of the same sort as the domainsort of the argument position the term r occupies in the paramodulant P. Then we use each member of a minimal and complete set of well-sorted unifiers for {r,z} as a coercion substitution, which applied to P yields a paramodulant of the ΣRP-calculus (and obviously the result is always a well-sorted clause).

Given the clause set of Figure 2.7, we paramodulate clauses 1(1) and 3(1), using the well-sorted unifier $\{l_1 \leftarrow b'\}$, and obtain $\{H(l_2)\}$. We let e_1 denote a fresh variable of sort E (i.e. the domainsort of H) and $\{\{l_2 \leftarrow b_1, e_1 \leftarrow b_1\},\{l_2 \leftarrow r_1, e_1 \leftarrow r_1\}\}$ is a minimal and complete set of well-sorted unifiers for $\{l_2, e_1\}$. The application of both coercion substitutions to $\{H(l_2)\}$ finally yields the many-sorted paramodulants $\{H(b_1)\}$ and $\{H(r_1)\}$.

With this definition of the ΣRP-calculus, we obtain the following refutation from the clause set of Figure 2.7:

(P1) $\{H(b_1)\}$; many-sorted paramodulant of 1(1) and 3(1)

(P2) $\{H(r_1)\}$; many-sorted paramodulant of 1(1) and 3(1)

(R3) $\{\neg H(r)\}$; many-sorted resolvent of P1(1) and 2(1)

(R4) □ ; many-sorted resolvent of P2(1) and R3(1)

Many-sorted Axiomatizations The principal idea of many-sorted logic is to represent certain truths by defining a sort hierarchy and a signature, instead of having these truths expressed explicitly by first-order formulas. Hence it is possible to incorporate certain deductions into the inference machinery of a calculus, instead of performing them explicitly by inference steps. In case of the ΣRP-calculus, this incorporation is realized (mainly) in the modified unification procedure.

But which kind of truths can be represented by a sort hierarchy and a signature ? We call statements of naive set theory, which can be defined in terms of the subsort order or of the signature, **sortal statements**. Figure 2.8. is a collection of sortal statements and their equivalent representations by the subsort order and the signature.

$P \subset Q$	$\leftrightarrow$	P is subsort of Q
$P \cap Q \neq \emptyset$	$\leftrightarrow$	some sort R is a subsort of P and Q
$q \in Q$	$\leftrightarrow$	q is a constant of sort Q
$f(P_1 \ldots P_n) \subset Q$	$\leftrightarrow$	f is a function with domainsorts $P_1,\ldots,P_n$ and rangesort Q

Figure 2.8 Sortal statements and their equivalent representation by the sort hierarchy and the signature

Obviously, the performance of a many-sorted logic increases with an increasing number of sortal statements in an axiomatization, because the number of explicit inferences decreases if we do not use sort literals. But unfortunately, we cannot express a **negative sortal statement**, i.e. the negation of a sortal statement, like $P \not\subset Q$ or $P \cap Q = \emptyset$, as a sortal statement. Consequently, we cannot represent each unary predicate symbol in an unsorted clause set by a sort symbol. Let us consider the following example:

<u>Example 2.4</u> [Chang and Lee 1973]

(i) Some patients like all doctors. (ii) No patient likes any quack. (iii) Therefore, no doctor is a quack.

We can represent the set of patients by a sort P. But if we represent the set of doctors and also the set of quacks by sorts, we are unable to axiomatize the conclusion (iii), because we cannot state that $D \cap Q = \emptyset$. Hence we have to choose sort D, say, to represent the set of doctors, and with a predicate Q(x) for "x is a quack" we can symbolize (iii) as $\forall x{:}D\ \neg Q(x)$.

However, there is a simple trick to have both doctors and quacks as sorts. Since we are interested in resolution based theorem proving, it suffices to show that the conjunction of statements (i) and (ii) with the negation of statement (iii) is unsatisfiable. But the negation of (iii), i.e. "there is a doctor who is a quack", can be represented within the sort hierarchy: we let X denote a sort symbol standing for the intersection of doctors and quacks, i.e. X is a subsort of D and of Q. By presence of X we have $D \cap Q \neq \emptyset$, i.e. a sortal statement for the negation of statement (iii). Figure 2.9 is a diagramm of this sort hierarchy.

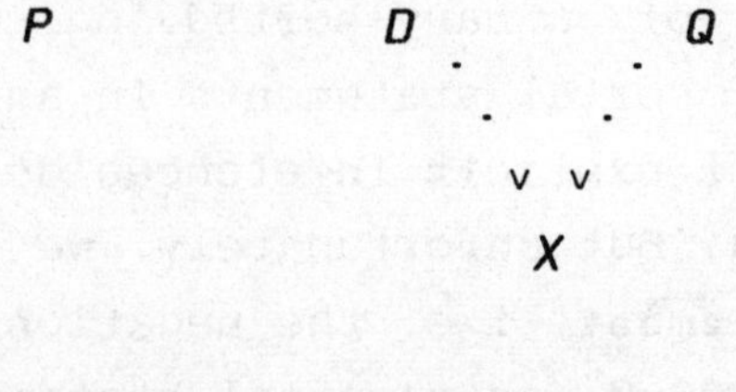

Figure 2.9 A sort hierarchy for Example 2.4

Using L(xy) for "x likes y", p as a constant of sort P, and d_1, p_1, q_1 and x_1 as variables of sort P,D,Q and X respectively, we obtain the following many-sorted clausal representation for Example 2.4:

(1) $\{L(p\ d_1)\}$ (2) $\{\neg L(p_1\ q_1\}$

With the well-sorted unifier $\{p_1 \leftarrow p, d_1 \leftarrow x_1, q_1 \leftarrow x_1\}$, we derive the empty clause with a single many-sorted resolution step. Note that without D and Q having a common subsort X, the variables d_1 and q_1 would not have a well-sorted unifier. For the corresponding unsorted axiomatization, we obtain the empty clause with 4 resolution steps from a set of 5 clauses with 8 literals [Chang and Lee 1973].

Obviously, we can use this method whenever the theorem to be shown can be represented as a disjunction of negative sortal statements. But it fails if we have a negative sortal statement in the axiomset of a problem. For problems of this kind, we take advantage of the fact that sortal statements can also be expressed by many-sorted first-order equations:

$P \subset Q$	$\leftrightarrow$	$\forall x{:}P\ \exists y{:}Q\ x \equiv y$
$P \cap Q \neq \emptyset$	$\leftrightarrow$	$\exists x{:}P\ \exists y{:}Q\ x \equiv y$
$q \in Q$	$\leftrightarrow$	$\exists y{:}Q\ q \equiv y$
$f(P_1 \ldots P_n) \subset Q$	$\leftrightarrow$	$\forall x_1, \ldots, x_n{:}P\ \exists y{:}Q\ f(x_1 \ldots x_n) \equiv y$

Figure 2.10 Sortal statements and their representation by many-sorted equations

Since we have these equations in our object language, we are also able to represent negative sortal statements, like

∀x:P ∀y:Q ¬x≡y stands for P∩Q=∅, without resorting to the use of sort literals. Let us illustrate the advantage of these axiomatizations with another example:

Example 2.5 [Chang and Lee 1973]
(i) No used-car dealer buys a used car for his family. (ii) Some people who buy used cars for their families are absolutely dishonest. (iii) Conclude that some absolutely dishonest people are not used-car dealers.

We let B,D,U and X be sorts standing for people, who buy used cars, absolutely dishonest people and used-car dealers. We define X as a subsort of B and of D, i.e. B∩D≠∅, and thus an axiomatization of statement (ii) by a sortal statement is obtained. We also define D as a subsort of U, i.e. D⊂U, and obtain an axiomatization of the negated statement (iii), viz. "all absolutely dishonest people are used-car dealers", by a sortal statement. Figure 2.11 is a diagram of this sort hierarchy.

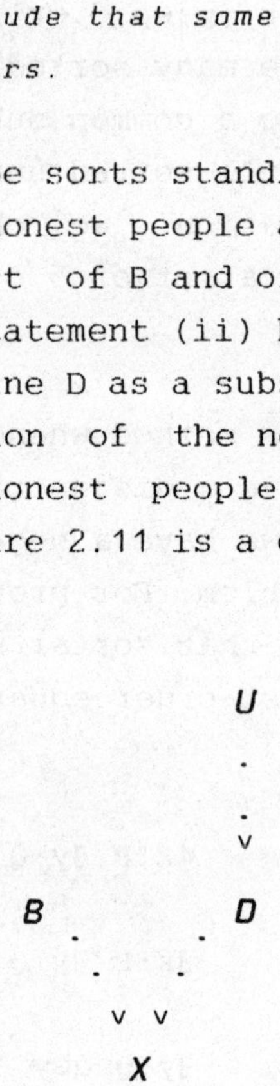

Figure 2.11 A sort hierarchy for Example 2.5

The remaining statement (i) is a negative sortal statement, viz. U∩B=∅, and is represented explicitly by the many-sorted unit clause

(1) $\{\neg u_1 \equiv b_1\}$

(compared to 4 clauses with 6 literals in an unsorted axiomatization). From (1), we obtain the empty clause by resolution with the reflexivity clause $\{z_1 \equiv z_1\}$, using the well-sorted unifier $\{z_1 \leftarrow x_1, u_1 \leftarrow x_1, b_1 \leftarrow x_1\}$, (where the sort of z_1 is any). Note that without U and B having a common subsort X, the variables u_1 and b_1 of sort U and B would not have a well-sorted unifier.

However, we use many-sorted equations not only for the representation of negative sortal statements. Since it may become necessary to verify sortal statements if the given sort hierarchy becomes very large or complex, we also need a means to represent non-negative sortal statements explicitly. Again, many-sorted equalities can be used to axiomatize sortal statements without sort literals:

Example 2.6

(i) Every man is mortal. (ii) Socrates is a man. (iii) Therefore Socrates is mortal.

Defining sort MAN as a subsort of sort MORTAL, and Socrates as a constant of sort MAN, we obtain an axiomatization of statements (i) and (ii) by sortal statements only. Hence we are unable to give an axiomatization for statement (iii) without sort literals, unless we use many-sorted equalities. For the example above, statement (iii) is represented as $\exists x{:}MORTAL\ socrates \equiv x$, which after negation yields the clause:

(1) $\{\neg socrates \equiv m_1\}$,

where m_1 is a variable of sort MORTAL. From (1), we obtain the empty clause by resolution with the reflexivity clause $\{z_1 \equiv z_1\}$, using the well-sorted unifier $\{z_1 \leftarrow socrates, m_1 \leftarrow socrates\}$.

3 Formal Preliminaries for the RP-Calculus

Here we present the basic notions for the RP-calculus and define the notation used subsequently. We assume that the reader is familiar with the basic notions of formal logic and automated theorem proving, cf. [Mendelson 1964, Loveland 1978, Chang and Lee 1973].

Syntactic Notions Given pairwise disjoint alphabets, the infinite set of variable symbols $\mathcal{V}$, the non-empty set of function symbols $\mathcal{F}$ and the non-empty set of predicate symbols $\mathcal{P}$, together with an arity-function for function and predicate symbols, we let T denote the set of all well formed terms over $\mathcal{V}$ and $\mathcal{F}$ and let AT denote the set of all well formed atoms over $\mathcal{V}$, $\mathcal{F}$ and $\mathcal{P}$. $\mathcal{C}$ stands for the set of all constant symbols, i.e. function symbols with arity 0.

A literal is an atom (also called a positive literal) or an expression of the form $\neg A$, where A is an atom (also called a negative literal). LIT denotes the set of all literals. A pair of literals is called complementary, if one of the literals is positive and the other is negative. Given a literal L, $|L|$ denotes the atom of L defined as $|L|=A$ iff $L=A$ or $L=\neg A$ where $A \in AT$. L^C denotes L's complement given by $L^C=A$ iff $L=\neg A$ and $L^C=\neg A$ iff $L=A$. The predicate letter of L is P iff $|L|=P$ or $|L|=P(t_1...t_n)$ for some $t_i \in T$. A clause is a finite set of literals and $\Box$ denotes the empty clause. The clause language $\mathcal{L}$ is the set of all clauses over $\mathcal{V}$, $\mathcal{F}$ and $\mathcal{P}$. For a term or literal t, $\mathcal{V}(t)$ denotes the set of all variable symbols in t. For a set D of terms or literals, $\mathcal{V}(D)$ is defined as $\cup_{t \in D} \mathcal{V}(t)$.

The subscript gr abbreviates ground, which stands for variable free, e.g. a ground term is a variable free term and $T_{gr}=\{t\in T| \mathcal{V}(t)=\emptyset\}$ is the set of all ground terms. AT_{gr}, LIT_{gr} and $\mathcal{L}_{gr}$ are defined in a similar way.

When concerned with equality reasoning, we use E as the syntactic equality sign and define $E\in\mathcal{P}$ with arity=2. S^E denotes the extension of the clause set S by all functionally-reflexive axioms [Wos and Robinson 1973]. The set of all equality atoms AT^E is defined as $AT^E=\{E(qr)|q,r\in T\}$.

Substitutions and Unifiers A mapping σ from $\mathcal{V}$ to T with $\sigma x=x$ almost everywhere is called a substitution. Substitutions are extended as endomorphisms to mappings from T to T. ε is the empty or identity substitution, i.e. $\varepsilon x=x$ for any $x\in\mathcal{V}$. For subsets D of T we define $\sigma D=\{\sigma t|t\in D\}$. The domain of σ, denoted $DOM(\sigma)$ is given by $\{x\in\mathcal{V}|\sigma x\neq x\}$ and the codomain (or range) $COD(\sigma)$ of σ is defined as $\sigma DOM(\sigma)$. Since $DOM(\sigma)$ is finite for each substitution σ, each substitution can be represented as (and defined by) a finite set of variable-term pairs $\{x_1\leftarrow t_1,\ldots,x_n\leftarrow t_n\}$, where $DOM(\sigma)=\{x_1,\ldots,x_n\}$ and $\sigma x_i=t_i$. SUB is the set of all substitutions.

The application σA of a substitution σ to an atom A=P or $A=P(t_1\ldots t_n)$ is given as P or $P(\sigma t_1\ldots\sigma t_n)$ respectively and $\sigma\neg A$ is defined as $\neg\sigma A$. For a set D of terms or literals we define $\sigma D=\{\sigma t|t\in D\}$.

The restriction $\sigma_{|V}$ of σ to a subset V of $\mathcal{V}$ is defined as $\sigma_{|V}x=\sigma x$, if $x\in V$, and otherwise as $\sigma_{|V}x=x$. A subset V of $\mathcal{V}$ induces an equivalence relation $=[\![V]\!]$ on SUB by $\sigma=\rho[\![V]\!]$ iff $\sigma x=\rho x$ for all $x\in V$. The following lemma is frequently used throughout the remainder:

Lemma 3.1

Let $\theta,\lambda\in SUB$, $t\in T$ and $V,W\subset\mathcal{V}$. Then

(1) $\theta=\lambda[\![V\cup W]\!]$ iff $\theta=\lambda[\![V]\!]$ and $\theta=\lambda[\![W]\!]$, and

(2) if $\mathcal{V}(t)\subset V$ and $\theta=\lambda[\![V]\!]$, then $\theta t=\lambda t$.

The quasi-ordering $\leq[\![V]\!]$ of **subsumption** in SUB over some $V\subset\mathcal{V}$ is given as $\sigma\leq\theta[\![V]\!]$ iff $\lambda\circ\sigma=\theta[\![V]\!]$ for some $\lambda\in SUB$ (where $\circ$ denotes functional composition). For $\sigma\leq\theta[\![V]\!]$ we say that σ is **more general than** θ in V. $\leq[\![V]\!]$ induces an equivalence relation $\equiv[\![V]\!]$ on SUB by $\sigma\equiv\rho[\![V]\!]$ iff $\sigma\leq\rho[\![V]\!]$ and $\rho\leq\sigma[\![V]\!]$. $[\![V]\!]$ is omitted in the above definitions if $V=\mathcal{V}$.

Given $V\subset\mathcal{V}$, a **renaming substitution** for V is a substitution ν satisfying (1) $DOM(\nu)=V$, (2) $COD(\nu)=\mathcal{V}\setminus V$ and (3) $|DOM(\nu)|=|COD(\nu)|$. A renaming substitution for a clause C or a set of clauses S is a renaming substitution for the set of variable symbols in C or in S.

We say that a substitution σ is a **ground substitution** iff $COD(\sigma)\subset T_{gr}$. SUB_{gr} denotes the set of all ground substitutions. For a clause C and a substitution σ, σC is called an **instance** of C. If $\sigma C\in\mathcal{L}_{gr}$, then σC is a **ground instance** of C.

Given a set D of terms or literals, we call a substitution θ a **unifier** of D and say that D is **unifiable** iff $|\theta D|\leq 1$. $\mathcal{U}(D)$ is the set of all unifiers of D. A unifier σ of D is called a **most general unifier** (or mgu for short) iff $\sigma\leq\theta$ for each $\theta\in\mathcal{U}(D)$. Note that $\mathcal{U}(\emptyset)=SUB$, hence $\varepsilon\in\mathcal{U}(\emptyset)$ and since $\lambda\circ\varepsilon=\lambda$ for all $\lambda\in SUB$ we have that ε is an mgu of $\emptyset$. The following lemma presents some elementary facts about unifiers and most general unifiers (cf. also [Eder 1985, Fages and Huet 1983, Herold 1983]):

Lemma 3.2

Let $D\subset T$ be unifiable. Then D has an mgu and for each mgu σ of D

(1) $DOM(\sigma)\cup\mathcal{V}(COD(\sigma))\subset\mathcal{V}(D)$,

(2) $\sigma\circ\sigma=\sigma$,

(3) $\theta=\theta\circ\sigma$ for all $\theta\in\mathcal{U}(D)$,

(4) $\sigma\equiv\tau$ for any mgu τ of D, and

(5) $|D|<\infty$.

Subterm Selectors A partial function $\delta:T\to T$ is an argument selector iff for some $k(\delta)\in\mathbb{N}$ and each term $t=f(t_1...t_n)$ with $k(\delta)\leq n$: $\delta(t)=t_{k(\delta)}$. SEL denotes the set of all argument selectors. The identity function on terms, an argument selector or a finite composition of argument selectors are called subterm selectors, or selectors for short. We let SEL* denote the set of all selectors and define SEL^+ as $SEL^*\setminus\{id_{|T}\}$. Selectors can be represented by finite strings of natural numbers called occurences or positions [Rosen 1973, Huet and Oppen 1980]: $id_{|T}$ is represented by the empty string and $\alpha=\delta_n\circ...\circ\delta_1\in SEL^+$ (where $\delta_i\in SEL$) by $\langle k(\delta_1)...k(\delta_n)\rangle$.

Each selector α induces a symmetric and transitive relation $\underset{\alpha}{\sim}$ on T by $q\underset{\alpha}{\sim}r$ iff $\alpha(q)\downarrow$, $\alpha(r)\downarrow$ and q differs from r at most on the subterms of q and r selected by α. $<$ is a partial order relation on SEL* defined as $\alpha<\beta$ iff $\alpha=\delta\circ\beta$ for some $\delta\in SEL^+$, i.e. α selects a subterm of the subterm selected by β.

A pair of selectors α and β are called weakly independent, denoted $\alpha\underline{\perp}\beta$, iff $\alpha\not<\beta$ and $\alpha\not>\beta$. α and β are strongly independent, written $\alpha\perp\beta$, iff $\alpha\not\leq\beta$ and $\alpha\not\geq\beta$, where $\alpha\leq\beta$ abbreviates $\alpha<\beta$ or $\alpha=\beta$.

For a given set of literals I, a pair of terms q,r and a selector α, the expression $q\underset{\alpha I}{\to}r$ is an abbreviation for: $q\underset{\alpha}{\sim}r$ and $E(\alpha(q)\alpha(r))\in I$. The following lemma is frequently used in subsequent chapters (cf. also [Rosen 1973]):

Lemma 3.3

Let $q,r\in T$, $\alpha,\beta\in SEL^*$, $I\subset LIT$ and $\sigma\in SUB$. Then

(1) if $q \underset{\alpha}{\sim} r$ and $\alpha(q)=\alpha(r)$, then $q=r$,

(2) $q \underset{\beta\circ\alpha}{\sim} r$ iff $\alpha(q) \underset{\beta}{\sim} \alpha(r)$ and $q \underset{\alpha}{\sim} r$,

(3) $q \underset{\beta\circ\alpha I}{\longrightarrow} r$ iff $\alpha(q) \underset{\beta I}{\rightarrow} \alpha(r)$ and $q \underset{\alpha}{\sim} r$,

(4) if $q \underset{\alpha}{\sim} r$, then $\sigma q \underset{\alpha}{\sim} \sigma r$,

(5) if $\alpha(q)\downarrow$, then $\sigma\alpha(q)=\alpha(\sigma q)$,

(6) if $\alpha\perp\beta$ and $q \underset{\beta}{\sim} r$, then $\alpha(q)=\alpha(r)$, and

(7) if $\alpha\perp\beta$ and $q \underset{\alpha I}{\rightarrow} q' \underset{\beta I}{\rightarrow} r$ for some $q'\in T$,
then $q \underset{\beta I}{\rightarrow} r' \underset{\alpha I}{\rightarrow} r$ for some $r'\in T$.

Selectors different from $id_{|T}$ are applied to atoms as to terms. For a literal L we define $\alpha(L)=\alpha(|L|)$. For a pair of literals L and K, we define $L \underset{\alpha}{\sim} K$ as for terms but with the proviso that either both literals have to be positive or both have to be negative. Additionally we define that for a literal L, $id_{|T}(L)\downarrow$ is always false. Then for a pair of literals L and K, $L \underset{\alpha I}{\rightarrow} K$ is defined as for terms and Lemma 3.3 holds for literals as well.

Term Rewriting A [ground] term rewriting system is a set of directed equations $\mathcal{R}=\{E(q_i r_i)\in AT^E_{gr} \mid i\in J\}$ where $J\subset\mathbb{N}$. We define $\Rightarrow_{\mathcal{R}}\subset T_{gr}\times T_{gr}$ by $q\Rightarrow_{\mathcal{R}} r$ iff $E(qr)\in\mathcal{R}$ and we use $\rightarrow_{\mathcal{R}}$ to denote the reduction relation associated with $\mathcal{R}$, i.e. $q\rightarrow_{\mathcal{R}} r$ iff $q \underset{\alpha\mathcal{R}}{\rightarrow} r$ for some $\alpha\in SEL^*$. We use the standard notation $\overset{+}{\rightarrow}_{\mathcal{R}}$ for the transitive closure of $\rightarrow_{\mathcal{R}}$ and $\overset{*}{\rightarrow}_{\mathcal{R}}$ for the reflexive closure of $\overset{+}{\rightarrow}_{\mathcal{R}}$. If for two ground terms q and r there exist a sequence $q_1,\ldots$ $\ldots,q_{n+1}\in T_{gr}$ and a sequence $\alpha_1,\ldots,\alpha_n\in SEL^*$ such that
$q=q_1 \underset{\alpha_1}{\longrightarrow}_{\mathcal{R}} q_2 \ldots q_n \underset{\alpha_n}{\longrightarrow}_{\mathcal{R}} q_{n+1}=r$, we call $q_1 \underset{\alpha_1}{\longrightarrow} \ldots \underset{\alpha_n}{\longrightarrow} q_{n+1}$ an $\mathcal{R}$-rewrite of r from q with length n.

$\mathcal{R}$ is called symmetric iff $\Rightarrow_{\mathcal{R}}$ is a symmetric relation. $\overset{*}{\rightarrow}_{\mathcal{R}}$ is an equivalence relation if $\mathcal{R}$ is symmetric.

We like to manipulate ground literals by a term rewriting system as well and extend $\rightarrow_{\mathbb{R}}$ in the obvious way, i.e. $L \rightarrow_{\mathbb{R}}^{+} K$ for a pair of ground literals L and K iff $L \rightarrow_{\alpha\mathbb{R}} K$ for some $\alpha \in SEL^{+}$.

For a set of literals I, the term rewriting system $\mathbb{R}(I)$ contained in I is defined as $\mathbb{R}(I) = I \cap AT_{gr}^{E}$.

Inference Rules and Deductions $Res(C,L,D,K,\sigma) = \sigma(C-L) \cup \sigma(D-K)$ is the resolvent of the clauses C and D upon the complementary literals $L \in C$ and $K \in D$, where σ is an mgu of $\{|L|,|K|\}$. A substitution σ factors a clause C and σC is a factor of C iff σ is an mgu of some subset of C.

$Par(C,L,D,E(qr),\alpha,\sigma) = \sigma(C-L) \cup \sigma(D-E(qr)) \cup \{\sigma K\}$ is a paramodulant of the clauses C and D upon $L \in C$ and $E(qr) \in D$ iff σ is an mgu of $\{\alpha(L),q\}$, $\sigma L \underset{\alpha}{\sim} \sigma K$ and $\alpha(\sigma K) = \sigma r$ (or σ is an mgu of $\{\alpha(L),r\}$, $\sigma L \underset{\alpha}{\sim} \sigma K$ and $\alpha(\sigma K) = \sigma q$), where σK is called the modulant literal [Loveland 1978].

Given a set S of clauses and a clause C, $S \vdash C$ denotes the existence of a deduction (in the RP-calculus) of C from S, i.e. there exists a list of clauses $\langle B_1,...,B_n \rangle$ such that $B_n = C$ and for each i with $1 \le i \le n$, $B_i = v_i C_i$ where $C_i \in S$ or C_i is a resolvent, paramodulant or factor of clauses preceding B_i in the list and v_i is a renaming substitution for $\{B_1,...,B_{i-1}\}$. The length of the deduction is given as the length n of the clause list $\langle B_1,...,B_n \rangle$. A refutation of S is a deduction of the empty clause from S. $S \vdash_R C$ is a deduction without paramodulation and $S \vdash_P C$ is a deduction without resolution.

Semantic Notions For a set of clauses S, S_{gr} denotes the set of all ground instances of the clauses in S. Computing S_{gr} for a given clause set S, we will agree that $\mathfrak{F}$ and $\mathfrak{P}$ are minimal, i.e. each symbol from $\mathfrak{F}$ and $\mathfrak{P}$ is used in at least one of the clauses of S. This guarantees that T_{gr} is the Herbrand Universe of S and AT_{gr} is the Herbrand Base of S [Loveland 1978], if we assume in addition that $\mathfrak{C}=\{c\}$ for the case that S contains no constant symbol at all.

A (possibly infinite) subset I of LIT_{gr} is called an interpretation iff for each $L\in I$, $L^c\notin I$. I is called reflexive iff for each $t\in T_{gr}$, $E(tt)\in I$. We say that I is E-closed iff for each $L\in I$ and each $K\in LIT_{gr}$: $K\in I$ whenever $L\underset{\alpha I}{\rightarrow}K$ for some $\alpha\in SEL^+$. A reflexive and E-closed interpretation is an E-interpretation. The following lemma is used constantly:

Lemma 3.4

Let $I\subset LIT_{gr}$ be E-closed. Then

(1) if I is reflexive, then $E(qr)\in I$ iff $E(rq)\in I$, and

(2) if $L\in I$, $K\in LIT_{gr}$ and $L\overset{*}{\rightarrow}_{\mathfrak{R}(I)}K$, then $K\in I$.

An interpretation I satisfies a ground clause C iff $I\cap C\neq\emptyset$. I satisfies a clause C iff I satisfies each ground instance σC of C. I satisfies a set of clauses S iff I satisfies each clause in S. In this case I is a model of S and S is satisfiable. If I is an E-interpretation, I E-satisfies S, I is an E-model of S and S is E-satisfiable.

4 Formal Preliminaries for the ΣRP-Calculus

In this chapter we present the basic notions and notations for the ΣRP-calculus.

Sorts and Signatures A *sort hierarchy* is a pair $(\mathcal{S},<_{\mathcal{S}})$ such that $\mathcal{S}$ is a non-empty set partially ordered by $<_{\mathcal{S}}$. The members of $\mathcal{S}$ are called *sort symbols* and $<_{\mathcal{S}}$ is the *subsort order* of $\mathcal{S}$. If $s_1<_{\mathcal{S}}s_2$ or $s_1=s_2$, one writes $s_1\leq_{\mathcal{S}}s_2$, and says s_1 *is a subsort* of s_2. The relation $s_1<_{\mathcal{S}}s_2$ is also writen $s_2>_{\mathcal{S}}s_1$ and similarly $s_1\leq_{\mathcal{S}}s_2$ is sometimes written $s_2\geq_{\mathcal{S}}s_1$. If $\mathcal{S}$ is known from the context we shall omit the indices, e.g. we write $<$ for $<_{\mathcal{S}}$.

A sort hierarchy $(\mathcal{S},<)$ is a *forest structure* iff $s_1\geq s\leq s_2$ implies $s_1\geq s_2$ or $s_1\leq s_2$ for all $s_1,s_2,s\in\mathcal{S}$. $(\mathcal{S},<)$ is a *tree structure* iff $(\mathcal{S},<)$ is a forest and $s\leq s\Delta$ for some $s\Delta\in\mathcal{S}$ and all $s\in\mathcal{S}$. $s'\in\mathcal{S}$ is *minimal* in $(\mathcal{S},<)$ iff $s\nless s'$ for all $s\in\mathcal{S}$.

For the remainder we shall assume that $\mathcal{S}$ is finite. Then $(\mathcal{S},<_{\mathcal{S}})$ satisfies the minimum condition and we can use the principle of *Noetherian induction*, cf. [Cohn 1981], in subsequent proofs.

For a given sort hierarchy $(\mathcal{S},<)$ let $\mathcal{S}^*$ be the set of all finite strings from $\mathcal{S}$, including the empty string e. For each $s\in\mathcal{S}$ and for each $w\in\mathcal{S}^*$ let $\mathcal{V}_s$ be a set of *variable symbols*, $\mathcal{F}_{w,s}$ a set of *function symbols* and $\mathcal{P}_w$ a set of *predicate symbols* such that all these sets are pairwise disjoint. Additionally we shall assume, that for each s which is *minimal* in $(\mathcal{S},<)$, $\mathcal{F}_{e,s}\neq\emptyset$. Then an *$\mathcal{S}$-sorted signature* Σ (for variable, function and predicate symbols) is a family $\Sigma_{w,s}$ of sets such that $\Sigma_{w,s}=\mathcal{V}_s\cup\mathcal{F}_{w,s}\cup\mathcal{P}_w$ for all $w\in\mathcal{S}^*$ and $s\in\mathcal{S}$. Setting $\mathcal{V}=\cup_{s\in\mathcal{S}}\mathcal{V}_s$, $\mathcal{F}=\cup_{w\in\mathcal{S}^*,s\in\mathcal{S}}\mathcal{F}_{w,s}$, $\mathcal{C}=\cup_{s\in\mathcal{S}}\mathcal{C}_s$ (where $\mathcal{C}_s=\mathcal{F}_{e,s}$) and $\mathcal{P}=\cup_{w\in\mathcal{S}^*}\mathcal{P}_w$, we define terms, atoms, literals etc. as in Chapter 3, where the arity of each $m \in \mathcal{F}_{w,s} \cup \mathcal{P}_w$ is given by the length of the string w.

If concerned with equality reasoning, we shall assume that $(\mathcal{S}, <_{\mathcal{S}})$ is a tree structure and we define $E \in \mathcal{P}_{s\Delta s\Delta}$.

Syntactic Notions For a variable or function symbol a, the rangesort [a] of a is s iff $a \in \mathcal{V}_s \cup \mathcal{F}_{w,s}$ (for some $w \in \mathcal{S}^*$). For a function or predicate symbol m, the ith domainsort $[m]_i$ of m is s(i) iff $m \in \mathcal{F}_{s(1)\ldots s(k),s} \cup \mathcal{P}_{s(1)\ldots s(k)}$, provided $1 \leq i \leq k$. The sort [t] of a term t is s iff $t \in \mathcal{V}_s \cup \mathcal{C}_s$ or $t = f(t_1 \ldots t_k)$ and [f]=s, i.e. [t] is determined by the outermost symbol of t.

For a term t and a selector α with $\alpha(t)\downarrow$ we define the α-maximal sort of t, denoted $[t]_\alpha$, by

(1) $[t]_\alpha = [t]$, if $\alpha = id_{|T}$,
(2) $[t]_\alpha = [f]_i$, if $t = f(t_1 \ldots t_n)$, $\alpha(t) = t_i$ and $\alpha \in SEL$, and
(3) $[t]_\alpha = [\delta(t)]_\beta$, if $\alpha = \beta \circ \delta$ with $\beta \in SEL$ and $\delta \in SEL^+$.

For $\alpha \in SEL^+$ we apply this definition to atoms as well and define for each literal L that $[L]_\alpha = [|L|]_\alpha$. The following lemma is easy to prove:

Lemma 4.1

Let $q, r \in T$, $\alpha, \beta \in SEL^*$ and $\sigma \in SUB$. Then
(1) if $\beta \in SEL^+$, then $[q]_{\beta\circ\alpha} = [\alpha(q)]_\beta$,
(2) if $q \notin \mathcal{V}$ and $\alpha(q)\downarrow$, then $[q]_\alpha = [\sigma q]_\alpha$,
(3) if $q \underset{\beta}{\sim} r$ and $\alpha \nleq \beta$, then $[\alpha(q)] = [\alpha(r)]$, and
(4) if $q \underset{\beta}{\sim} r$ and $\alpha \not\leq \beta$, then $[q]_\alpha = [r]_\alpha$.

Given an $\mathcal{S}$-sorted signature Σ, a term t is called a well-sorted term or a Σ-term iff $[\alpha(t)] \leq [t]_\alpha$ for each selector α with $\alpha(t)\downarrow$. We say that an atom A is well-sorted or A is a Σ-atom iff $A \in \mathcal{P}_e$ or $[\alpha(A)] \leq [A]_\alpha$ for each selector α with $\alpha(A)\downarrow$. T_Σ denotes the set of all Σ-terms, AT_Σ denotes the set of all Σ-atoms and LIT_Σ is the set of all Σ-literals. Later we shall frequently use the following lemmata:

<u>Lemma 4.2</u>

Let $q \in T_\Sigma$, $r \in T$ and $\alpha \in SEL^*$.
If $q \underset{\alpha}{\sim} r$, $\alpha(r) \in T_\Sigma$ and $[\alpha(r)] \leq [r]_\alpha$, then $r \in T_\Sigma$.

<u>Lemma 4.3</u>

For all $t \in T$, $t \in T_\Sigma$ iff
(1) $t \in \mathcal{V}_s \cup \mathcal{C}_s$ for some $s \in \mathcal{S}$, or
(2) there exist $s, s(1), \ldots, s(n) \in \mathcal{S}$, $f \in \Sigma_{s(1)\ldots s(n),s}$ and $q_1, \ldots, q_n \in T_\Sigma$ such that $t = f(q_1 \ldots q_n)$ and $[q_i] \leq s(i)$ for each i with $1 \leq i \leq n$.

By Lemma 4.3, a term t is well-sorted iff each argument position i of each function symbol f in t is occupied by a term, the sort of which is less than or equal to the sort which is stipulated for the argument position i and the function symbol f in the signature. Obviously these lemmata hold for literals as well.

A **well-sorted** clause or a Σ-**clause** is a finite set of Σ-literals. The **many-sorted clause language** $\mathcal{L}_\Sigma$ is the set of all Σ-clauses. $T_{\Sigma gr}$ denotes the set of all variable free Σ-terms. $AT_{\Sigma gr}$, $LIT_{\Sigma gr}$ and $\mathcal{L}_{\Sigma gr}$ are defined in a similar way.

Note that these notions are consistent with the standard terminology: for $<_\mathcal{S} = \emptyset$, i.e. $s_1 \leq_\mathcal{S} s_2$ iff $s_1 = s_2$, T_Σ is the **free Σ-algebra generated** by $\mathcal{V}$ denoted $T(\Sigma \cup \mathcal{V})$ in [Huet and Oppen 1980] or $T_\Sigma(\mathcal{V})$ in [Goguen et al. 1978]. T_Σ is the Σ-**word algebra** on $\mathcal{V}$, denoted $W_\Sigma(\mathcal{V})$ in [Cohn 1981], if $\mathcal{S}$ is a singleton. The difference is that for a sequence $q_1, \ldots, q_k$ of Σ-terms and some $f \in \mathcal{F}_{s(1)\ldots s(k),s}$, $f(q_1 \ldots q_k) \in T_\Sigma$ entails $[q_i] \leq s(i)$, whereas in the unordered (and standard) case $[q_i] = s(i)$ is required.

Substitutions and Unifiers A substitution σ satisfying $\sigma T_\Sigma \subset T_\Sigma$ is called a well-sorted substitution or a Σ-substitution. A more useful characterization is $[\sigma x] \leq [x]$ and $\sigma x \in T_\Sigma$ for all $x \in \mathcal{V}$ and it can be proved that both definitions are equivalent (see Lemma 7.1). SUB_Σ is the set of all Σ-substitutions. Again these notions are in accordance with the many-sorted unordered case, where a substitution is defined as a sort-preserving Σ-endomorphism from T_Σ to T_Σ [Huet and Oppen 1980], thus guaranteeing $[\sigma x]=[x]$ and $\sigma x \in T_\Sigma$ for each $x \in \mathcal{V}$.

The quasi-ordering $\leq_\Sigma[\![V]\!]$ of Σ-subsumption in SUB_Σ over some $V \subset \mathcal{V}$ is given as $\sigma \leq_\Sigma \theta[\![V]\!]$ iff $\lambda \circ \sigma = \theta[\![V]\!]$ for some $\lambda \in SUB_\Sigma$ and the corresponding equivalence relation $\equiv_\Sigma[\![V]\!]$ is defined by $\sigma \equiv_\Sigma \rho[\![V]\!]$ iff $\sigma \leq_\Sigma \rho[\![V]\!]$ and $\rho \leq_\Sigma \sigma[\![V]\!]$. Again $[\![V]\!]$ is omitted if $V=\mathcal{V}$. The following lemma is easily shown:

Lemma 4.4
Let $\theta, \sigma \in SUB_\Sigma$ and $\lambda \in SUB$. Then
(1) $\theta \circ \sigma \in SUB_\Sigma$,
(2) if $\theta \leq \sigma$, then $\theta \leq_\Sigma \sigma$,
(3) $\sigma(LIT_\Sigma) \subset LIT_\Sigma$, and
(4) $\sigma(\mathcal{L}_\Sigma) \subset \mathcal{L}_\Sigma$.

A Σ-renaming substitution ν for a set D of variables, literals or clauses is a renaming substitution for D such that $[\nu x]=[x]$ for each $x \in \mathcal{V}$, and obviously $\nu \in SUB_\Sigma$. A Σ-ground substitution σ is a Σ-substitution with $COD(\sigma) \subset T_{\Sigma gr}$. $SUB_{\Sigma gr}$ denotes the set of all Σ-ground substitutions.

For a Σ-clause C and a Σ-substitution σ, σC is called a Σ-instance of C. If $\sigma C \in \mathcal{L}_{\Sigma gr}$, then σC is a Σ-ground instance of C.

Given a set D of Σ-terms or Σ-literals and a substitution θ, θ is a Σ-unifier of D and D is Σ-unifiable by θ iff θ is well-sorted and unifies D. $\mathcal{U}_\Sigma(D)=\mathcal{U}(D) \cap SUB_\Sigma$ is the set of all Σ-unifiers of D. σ is a **well-sorted most general unifier** or a Σ-mgu of D iff $\sigma \in SUB_{\Sigma}$ is an mgu of D.

Unfortunately the notion of a most general unifier is too strong to provide us with all the necessary concepts when unification is generalized to the many-sorted case (with $<_{\varphi} \neq \emptyset$). The situation is similar (but not identical) to that of unification under equational theories, where the notion of an mgu is replaced by the concept of complete and minimal sets of unifiers, cf. [Plotkin 1972, Fage and Huet 1983, Siekmann 1984]. According to the standard terminology we define:

$U \subset U_{\Sigma}(D)$ is a **complete** set of Σ-unifiers of some $D \subset T_{\Sigma}$ (or $D \subset LIT_{\Sigma}$) iff for each $\theta \in U_{\Sigma}(D)$ there is some $\sigma \in U$ such that $\sigma \leq_{\Sigma} \theta [\![\mathcal{V}(D)]\!]$. $U \subset U_{\Sigma}(D)$ is **minimal** iff $\sigma \leq_{\Sigma} \rho [\![\mathcal{V}(D)]\!]$ implies $\sigma = \rho$ for all $\sigma, \rho \in U$. Each complete and minimal set of Σ-unifiers is unique up to variable renamings, i.e. for any complete and minimal pair $U_1, U_2 \subset U_{\Sigma}(D)$ we have: for each $\sigma \in U_1$ there exists exactly one $\rho \in U_2$ such that $\sigma \equiv_{\Sigma} \rho [\![\mathcal{V}(D)]\!]$ and therefore $|U_1| = |U_2|$ (the proof is the same as in the unsorted case, cf. [Fage and Huet 1983]). We let $mcU_{\Sigma}(D)$ denote an arbitrary minimal and complete subset of $U_{\Sigma}(D)$, if such a set does exist at all, and define $mcU_{\Sigma}(D) = \emptyset$ otherwise.

Although these notions are similar to the notions used for unification under equational theories, there are essential differences. Equality $=$ is generalized to equality $=_E$ modulo some equational theory E but the notion of substitutions is not affected when equational reasoning is introduced to unification theory. In many-sorted theories substitutions are restricted to well-sorted ones but equality remains unchanged. Hence $U_{\Sigma}(D) \subset U(D) \subset U_E(D)$ for each $D \subset T_{\Sigma}$ and an equational theory E.

Term Rewriting For a (ground) term rewriting system $\mathcal{R}$ we define the Σ-reduction relation $\rightarrow_{\Sigma\mathcal{R}}$ associated with $\mathcal{R}$ and a signature Σ by $\rightarrow_{\Sigma\mathcal{R}_*} = \rightarrow_{\mathcal{R}} \cap (T_{\Sigma gr} \times T_{\Sigma gr})$. $\overset{+}{\rightarrow}_{\Sigma\mathcal{R}}$ is the transitive closure of $\rightarrow_{\Sigma\mathcal{R}}$ and $\overset{*}{\rightarrow}_{\Sigma\mathcal{R}}$ is the reflexive closure of $\overset{+}{\rightarrow}_{\Sigma\mathcal{R}}$. $\mathcal{R}$ is a Σ-maximal term rewriting system iff $\Rightarrow_{\mathcal{R}} = \overset{+}{\rightarrow}_{\Sigma\mathcal{R}}$. Note that in general a Σ-maximal term rewriting system is infinite. The following lemma is frequently used throughout the remainder:

<u>Lemma 4.5</u>

Let $\mathcal{R}$ be a Σ-maximal term rewriting system, $q, r \in T_{gr}$, $t \in \{q, r\}$ and $\alpha, \beta \in SEL^*$. Then

(1) if $q \Rightarrow_{\mathcal{R}} r$, then $t \in T_{\Sigma gr}$, and

(2) if $q \underset{\beta\mathcal{R}}{\Rightarrow} r$, $\alpha \overset{?}{<} \beta$ and $\alpha(t)\downarrow$, then $[\alpha(t)] \leq [t]_{\alpha}$.

Inference Rules and Deductions $Res_{\Sigma}(C, L, D, K, \sigma) = \sigma(C-L) \cup \sigma(D-K)$ is a Σ-resolvent of the Σ-clauses C and D upon the literals $L \in D$ and $K \in D$ iff L and K are complementary and $\sigma \in mcu_{\Sigma}(\{|L|, |K|\})$. A substitution σ Σ-factors a Σ-clause C and σC is a Σ-factor of C iff $\sigma \in mcu_{\Sigma}(D)$ for some $D \subset C$. Since we apply only Σ-substitutions to well-sorted clauses, each Σ-resolvent and each Σ-factor is a well-sorted clause.

$Par_{\Sigma}(C, L, D, E(qr), \alpha, \nu \circ \sigma) = \nu\sigma(C-L) \cup \nu\sigma(D-E(qr)) \cup \{\nu\sigma K\}$ is a Σ-paramodulant of the Σ-clauses C and D upon $L \in C$ and $E(qr) \in D$ iff

(1) $\sigma \in mcu_{\Sigma}(\{\alpha(L), t_1\})$,

(2) $\sigma L \underset{\alpha}{\sim} \sigma K$ and $\alpha(\sigma K) = \sigma t_2$, and

(3) $\nu \in mcu_{\Sigma}(\{z, \sigma t_2\})$,

where $t_1 = q$, $t_2 = r$ or $t_1 = r$, $t_2 = q$ and $z \in \mathcal{V}_{[L]\alpha} \setminus \mathcal{V}(C \cup D)$. The substitution ν, called a coercion substitution, guarantees that each Σ-paramodulant is a well-sorted clause (cf. Chapters 6 and 8). It can be thought of as a kind of coercion which is necessary in programming languages with types, e.g. PASCAL or ADA, whenever the types of a variable and an expression do not agree in an assignment.

Given a set of Σ-clauses S, $S\vdash_{\Sigma}C$ denotes the existence of a Σ-deduction of C from S, i.e. there exists a list of Σ-clauses $\langle B_1,\ldots,B_n\rangle$ such that B_n=C and for each i with $1\leq i\leq n$, $B_i = \nu_i C_i$ where $C_i \in S$ or C_i is a Σ-resolvent, Σ-factor or a Σ-paramodulant of the clauses preceeding B_i in the list and ν_i is a Σ-renaming substitution for $\{B_1,\ldots,B_{i-1}\}$. A Σ-refutation is a Σ-deduction of the empty clause. $S\vdash_{\Sigma R}C$ denotes a Σ-deduction without Σ-paramodulants and $S\vdash_{\Sigma P}C$ is a Σ-deduction without Σ-resolution.

Semantic Notions Given a set of Σ-clauses S, $S_{\Sigma gr}$ denotes the set of all Σ-ground instances of the Σ-clauses in S. An interpretation I Σ-satisfies a Σ-clause C iff I satisfies each Σ-ground instance σC of C. I Σ-satisfies a set of Σ-clauses S iff I Σ-satisfies each clause in S. In this case, I is a Σ-model of S and S is Σ-satisfiable. If in addition I is an E-interpretation, then I ΣE-satisfies S, I is a ΣE-model of S and S is ΣE-satisfiable. The following lemma presents some elementary facts about Σ(E)-(un)satisfiability:

Lemma 4.6

Let $S\subset\mathscr{L}_{\Sigma}$ and $I\subset LIT_{gr}$ be an interpretation. Then

(1) S is Σ-unsatisfiable iff $S_{\Sigma gr}$ is unsatisfiable,

(2) S is ΣE-unsatisfiable iff $S_{\Sigma gr}$ is E-unsatisfiable,

(3) $S_{\Sigma gr}\subset S_{gr}$, and

(4) if I Σ-satisfies S, then $I\cap LIT_{\Sigma gr}$ Σ-satisfies S.

Note that 4.6 (4) in general does not hold for ΣE-satisfiability because there exist E-interpretations I such that $I\cap LIT_{\Sigma gr}$ is neither reflexive nor E-closed and hence is not an E-interpretation.

Throughout the remainder $(\mathcal{S},<)$ is a partially ordered set of sort-symbols, Σ is some $\mathcal{S}$-sorted signature and S stands for any variable disjoint set of Σ-clauses.

5 Many-Sorted Term Rewriting

E-interpretations and E-models can easily be dealt with formally by means of ground term rewriting systems if the E-closure of an interpretation I is defined by the reduction relation $\rightarrow_{\mathcal{R}(I)}$.

Ground term rewriting systems and their Σ-reduction relation are particulary significant for the completeness of Σ-paramodulation in the ground case. As we shall see in Chapter 6, the completeness of Σ-paramodulation essentially depends on the fact that the relations $\overset{+}{\rightarrow}_{\mathcal{R}}$ and $\overset{+}{\rightarrow}_{\Sigma\mathcal{R}}$ correspond on $T_{\Sigma gr} \times T_{\Sigma gr}$ for certain ground term rewriting systems $\mathcal{R}$. What we are looking for is a criterion for ground term rewriting systems $\mathcal{R}$ which is sufficient for

$$\overset{+}{\rightarrow}_{\mathcal{R}} \cap (T_{\Sigma gr} \times T_{\Sigma gr}) = \overset{+}{\rightarrow}_{\Sigma\mathcal{R}}.$$

Here the problem is that a rewrite $q \overset{+}{\rightarrow}_{\mathcal{R}} r$ with $q, r \in T_{\Sigma gr}$ generally contains terms as intermediate results, which are not well-sorted. This is shown in the following example:

<u>Example 5.1</u>
Let $\mathcal{S}=\{A,B\}$, $f \in \mathcal{F}_{A,B}$, $g \in \mathcal{F}_{B,B}$, $a,b \in \mathcal{F}_{e,A}$ and $c,d \in \mathcal{F}_{e,B}$. Then $\mathcal{R}=\{$(i) $a \Rightarrow g(c)$, (ii) $g(c) \Rightarrow g(d)$, (iii) $g(d) \Rightarrow b\}$ is a term rewriting system with $f(a) \overset{+}{\rightarrow}_{\mathcal{R}} f(b)$, because

$$(*)\ f(a) \overset{(i)}{\rightarrow} f(g(c)) \overset{(ii)}{\rightarrow} f(g(d)) \overset{(iii)}{\rightarrow} f(b)$$

is an $\mathcal{R}$-rewrite. As can easily be seen, this is the only $\mathcal{R}$-rewrite of f(b) from f(a). With f(g(c)) and f(g(d)), however, this $\mathcal{R}$-rewrite contains terms which are not well-sorted and therefore $f(a) \overset{+}{\not\rightarrow}_{\Sigma\mathcal{R}} f(b)$. ⊠

But which demand would $\mathcal{R}$ now have to satisfy so that f(a) $\overset{+}{\rightarrow}_{\Sigma\mathcal{R}}$ f(b) is still possible?

As can be seen in the $\mathcal{R}$-rewrite (*) of Example 5.1, starting with the well-sorted term f(a), at first an ill-sorted reduction is made which, however, is then "corrected" during the $\mathcal{R}$-rewrite as in the end the well-sorted term f(b) is obtained. If the directed equation (i) by which we leave $T_{\Sigma gr}$ in (*), is combined with the directed equation (iii), which again leads back to $T_{\Sigma gr}$ in (*), to form a new directed equation (iv) a⇒b, and (iv) is added to $\mathcal{R}$, we obtain the new $\mathcal{R}$-rewrite f(a)$\xrightarrow{(iv)}$f(b) and hence f(a)$\overset{+}{\longrightarrow}_{\Sigma\mathcal{R}}$f(b).

This example suggests to demand the transitivity of a ground term rewriting system $\mathcal{R}$, i.e. $\Rightarrow_{\mathcal{R}} = \overset{+}{\Rightarrow}_{\mathcal{R}}$ to guarantee $\overset{+}{\rightarrow}_{\mathcal{R}} = \overset{+}{\rightarrow}_{\Sigma\mathcal{R}}$ on $T_{\Sigma gr} \times T_{\Sigma gr}$.

The following example illustrates however that the transitivity alone is not yet sufficient:

<u>Example 5.2</u>

Let $\mathcal{S}$ and $\mathcal{R}$ be the same as in Example 5.1 except that $\mathcal{R}$ contains the directed equation (ii') c⇒d instead of (ii). As the rewrite (*) from Example 5.1 is also here the only $\mathcal{R}$-rewrite of f(b) from f(a) and (*) contains ill-sorted terms then f(a)$\overset{+}{\not\longrightarrow}_{\Sigma\mathcal{R}}$ f(b) still holds even though $\Rightarrow_{\mathcal{R}} = \overset{+}{\Rightarrow}_{\mathcal{R}}$ is now satisfied. ⊠

The problem here is that the transitivity cannot be taken advantage of as only the subterm c of the term g(c), which makes f(g(c)) ill-sorted, and not g(c) is directly replaced.

It must therefore be demanded that $\mathcal{R}$ with c⇒d also contains g(c)⇒g(d). If additionally $\mathcal{R}$ is transitive we obtain a$\Rightarrow_{\mathcal{R}}$b as in the previous case and therefore f(a) $\overset{+}{\rightarrow}_{\Sigma\mathcal{R}}$ f(b).

For a ground term rewriting system $\mathcal{R}$ it is therefore necessary in general that $\mathcal{R}$ is transitive and closed w.r.t. $\rightarrow_{\Sigma\mathcal{R}}$, i.e. $\Rightarrow_{\mathcal{R}} = \overset{+}{\Rightarrow}_{\mathcal{R}} \cup \rightarrow_{\Sigma\mathcal{R}}$ is a sufficient condition to guarantee that $\overset{+}{\rightarrow}_{\mathcal{R}} \cap (T_{\Sigma gr} \times T_{\Sigma gr}) = \overset{+}{\rightarrow}_{\Sigma\mathcal{R}}$.

As we presume here that $\Rightarrow_{\mathcal{R}} \subset T_{\Sigma gr} \times T_{\Sigma gr}$ this condition, as can easily be shown, is equivalent to $\Rightarrow_{\mathcal{R}} = \overset{+}{\rightarrow}_{\Sigma\mathcal{R}}$, i.e. to the Σ-maximality of $\mathcal{R}$.

In conclusion we arrive at the following main theorem which is to be proven in this chapter:

<u>Σ-Rewrite Theorem</u> If $\mathcal{R}$ is a Σ-maximal ground term rewriting system, then $\overset{+}{\rightarrow}_{\mathcal{R}} \cap (T_{\Sigma gr} \times T_{\Sigma gr}) = \overset{+}{\rightarrow}_{\Sigma\mathcal{R}}$.

The main difficulty in proving this theorem is the proof of

$$\overset{+}{\rightarrow}_{\mathcal{R}} \cap (T_{\Sigma gr} \times T_{\Sigma gr}) \subset \overset{+}{\rightarrow}_{\Sigma\mathcal{R}}.$$

This will be shown by induction upon the length n of an $\mathcal{R}$-rewrite

$$(1)\quad q_1 \underset{\alpha_1}{\longrightarrow} q_2 \;\ldots\; q_n \underset{\alpha_n}{\longrightarrow} q_{n+1}, \qquad q_1, q_{n+1} \in T_{\Sigma gr}$$

A constructive method is presented by which from each $\mathcal{R}$-rewrite (1) which satisfies $\{q_2, \ldots, q_n\} \not\subset T_{\Sigma gr}$ an $\mathcal{R}$-rewrite $q_1 \rightarrow r_2 \ldots r_{n-1} \rightarrow q_{n+1}$ is obtained which is shorter than the given $\mathcal{R}$-rewrite. In this way all the terms $q_i \notin T_{\Sigma gr}$ from the given $\mathcal{R}$-rewrite are eliminated step by step and finally an $\mathcal{R}$-rewrite of q_{n+1} from q_1 is obtained, in which all the terms are well-sorted, i.e. for which $q_1 \overset{+}{\rightarrow}_{\Sigma\mathcal{R}} q_{n+1}$ holds true.

The constructive method can be outlined as follows:

First an $\mathcal{R}$-rewrite

(2) $q_{i-1} \xrightarrow[\alpha_{i-1}]{} q_i \ldots q_{j-1} \xrightarrow[\alpha_{j-1}]{} q_j$, $(2 \leq i < j \leq n+1)$

is selected from (1) such that $\alpha_{i-1} = \alpha_{j-1}$ and all terms $\alpha_{i-1}(q_h)$ with $h \in \{i, \ldots, j-1\}$ are well-sorted, but do not satisfy the signature in the position α_{i-1} for all q_h, i.e. $[\alpha_{i-1}(q_h)] \not\leq [q_h]\alpha_{i-1}$. The existence of such an $\mathcal{R}$-rewrite is guaranteed by the Σ-Rewrite Lemma (5.1).

With the the Shift-Up Lemma (5.2) we then show that (2) is still an $\mathcal{R}$-rewrite, if each selector α_h $(i \leq h \leq j-1)$, with $\alpha_h \geq \alpha_{i-1}$, is replaced in (2) by α_{i-1}. In Example 5.2 this corresponds to the change from $f(g(c)) \rightarrow_{\mathcal{R}} f(g(d))$ using $c \Rightarrow_{\mathcal{R}} d$ to $f(g(c)) \rightarrow_{\mathcal{R}} f(g(d))$ using $g(c) \Rightarrow_{\mathcal{R}} g(d)$, which is always possible in a ground term rewriting system $\mathcal{R}$ with $\rightarrow_{\Sigma\mathcal{R}} \subset \Rightarrow_{\mathcal{R}}$.

The replacement of the selectors α_h by α_{i-1} in (2) guarantees that $\alpha_h \perp \alpha_{j-1}$ holds true. Hence we can apply the Shift Left Lemma (5.3), according to which there is an $\mathcal{R}$-rewrite to (2), in which the sequence of the reductions is shifted in such a way that the first reduction is now carried out using the selector α_{j-1}, i.e. there is an $\mathcal{R}$-rewrite of the form

(3) $q_{i-1} \xrightarrow[\alpha_{j-1}]{} r_i \xrightarrow[\alpha_{i-1}]{} r_{i+1} \ldots r_{j-1} \xrightarrow[\alpha_{j-2}]{} q_j$.

Since $\alpha_{j-1} = \alpha_{i-1}$, (3) can be shortened by the Reduction Lemma (5.4) yielding an $\mathcal{R}$-rewrite

(4) $q_{i-1} \xrightarrow[\alpha_{i-1}]{} r_{i+1} \ \ldots \ r_{j-1} \xrightarrow[\alpha_{j-2}]{} q_j$

In this way an $\mathcal{R}$-rewrite

(5) $q_1 \underset{\alpha_1}{\longrightarrow} q_2 \ldots q_{i-1} \underset{\alpha_{i-1}}{\longrightarrow} r_{i+1} \ldots r_{j-1} \underset{\alpha_{j-2}}{\longrightarrow} q_j \ldots q_n \underset{\alpha_n}{\longrightarrow} q_{n+1}$

of length n-1 has been found, i.e. an $\mathcal{R}$-rewrite of q_{n+1} from q_1 which is shorter than the given $\mathcal{R}$-rewrite (1).

In the remainder of this chapter we deal with the proofs of the technical lemmata mentioned above which are used to prove the main result of this chapter, viz. the Σ-Rewrite Theorem.

<u>Lemma 5.1</u> (Σ-Rewrite Lemma)
Let $\mathcal{R}$ be a Σ-maximal ground term rewriting system and

(1) $q_1 \underset{\alpha_1}{\longrightarrow} q_2 \ldots q_n \underset{\alpha_n}{\longrightarrow} q_{n+1}$, $n \geq 1$,

be an $\mathcal{R}$-rewrite such that $q_1, q_{n+1} \in T_{\Sigma gr}$. If $\{q_2, \ldots, q_n\} \not\subset T_{\Sigma gr}$ then there exist indices i and j with $2 \leq i < j \leq n+1$, such that for each h with $i \leq h \leq j-1$

(2) $\alpha_{j-1} = \alpha_{i-1}$,

(3) $\alpha_{i-1}(q_h) \in T_{\Sigma gr}$, and

(4) $[\alpha_{i-1}(q_h)] \not\leq [q_h]_{\alpha_{i-1}}$.

Proof Since $\{q_2, \ldots, q_n\} \not\subset T_{\Sigma gr}$, we know that there exist some $q \in \{q_2, \ldots, q_n\}$ and some $\beta \in SEL^{\ddagger}$ such that $\beta(q)\downarrow$ but $[\beta(q)] \not\leq [q]_\beta$. Among those selectors β we can choose a selector α which is minimal w.r.t. $\overset{\circ}{<}$. Let q_m be a term of the $\mathcal{R}$-rewrite (1) with

(5) $\alpha(q_m)\downarrow$ and $[\alpha(q_m)] \not\leq [q_m]_\alpha$, (where $2 \leq m \leq n$).

Now starting on q_m we move left in the $\mathcal{R}$-rewrite until we find the first term q_{i-1} with

(6) $[\alpha(q_{i-1})] \leq [q_{i-1}]_\alpha$, if $\alpha(q_{i-1})\downarrow$ (where $i-1 < m$)

and starting again on q_m we move right in the ℛ-rewrite until we find the first term q_j with

(7) $[\alpha(q_j)] \leq [q_j]_\alpha$, if $\alpha(q_j)\downarrow$ (where $m<j$).

The existence of q_{i-1} is guaranteed, because the leftmost term q_1 in the ℛ-rewrite is a Σ-term and hence satisfies $[\alpha(q_1)] \leq [q_1]_\alpha$, if $\alpha(q_1)\downarrow$. By the same argument the rightmost term q_{n+1} in the ℛ-rewrite guarantees the existence of q_j. Since q_{i-1} is the first term to the left of q_m which satisfies (6), we know that $\alpha(q_i)\downarrow$ and $[\alpha(q_i)] \nleq [q_i]_\alpha$. The following cases show that $\alpha(q_{i-1})\downarrow$ also holds true:

Case (i) $\alpha \mathring{<} \alpha_{i-1}$: Using $q_{i-1} \xrightarrow[\alpha_{i-1}]{}_{ℛ} q_i$ and $\alpha(q_i)\downarrow$ we infer by Lemma 4.5 (2) that $[\alpha(q_i)] \leq [q_i]_\alpha$. ∇ ∅

Case (ii) $\alpha \mathring{\geq} \alpha_{i-1}$: From $q_{i-1} \xrightarrow[\alpha_{i-1}]{}_{ℛ} q_i$ we infer that $\alpha_{i-1}(q_{i-1})\downarrow$, hence $\alpha(q_{i-1})\downarrow$. ∅

Case (iii) $\alpha \perp \alpha_{i-1}$: From $q_{i-1} \xrightarrow[\alpha_{i-1}]{}_{ℛ} q_i$ we infer by Lemma 3.3 (6) that $\alpha(q_{i-1})=\alpha(q_i)$ and with $\alpha(q_i)\downarrow$ we obtain $\alpha(q_{i-1})\downarrow$. ∅

Hence we have proved that $\alpha(q_{i-1})\downarrow$ and using (6) we can write

(8) $\alpha(q_{i-1})\downarrow$ and $[\alpha(q_{i-1})] \leq [q_{i-1}]_\alpha$ (where $i-1<m$).

By a similar argument we prove that $\alpha(q_j)\downarrow$ and we obtain with (7)

(9) $\alpha(q_j)\downarrow$ and $[\alpha(q_j)] \leq [q_j]_\alpha$ (where $m<j$).

With q_{i-1} and q_j we have found terms in the ℛ-rewrite (1) which satisfy the signature on α. The following cases show that q_i and q_{j-1} are the terms in (1) such that we leave $T_{\Sigma gr}$ with

$q_{i-1} \to_{\mathcal{R}} q_i$ on α and do not return to $T_{\Sigma gr}$ until $q_{j-1} \to_{\mathcal{R}} q_j$ (cf. Example 5.1). Otherwise there must exist some h' with $i \le h' \le j-1$ such that $[\alpha(q_{h'})] \le [q_{h'}]_\alpha$, if $\alpha(q_{h'})\downarrow$:

Case (i) $i \le h' < m$: Then $q_{h'}$ instead of q_{i-1} is the first term to the left of q_m with $[\alpha(q_{h'})] \le [q_{h'}]_\alpha$, if $\alpha(q_{h'})\downarrow$. ∇ ∅

Case (ii) $h' = m$: This case is impossible by (5). ∇ ∅

Case (iii) $m < h' \le j-1$: Then $q_{h'}$ instead of q_j is the first term to the right of q_m with $[\alpha(q_{h'})] \le [q_{h'}]_\alpha$, if $\alpha(q_{h'})\downarrow$. ∇ ∅

Thus we have proved

(10) $\alpha(q_h)\downarrow$ and $[\alpha(q_h)] \not\le [q_h]_\alpha$, for each h with $i \le h \le j-1$,

i.e. all terms in (1) between the indices i and j-1 do not satisfy the signature on α. But all subterms of the terms selected by α are well-sorted because otherwise $\alpha(q_{h'}) \notin T_{\Sigma gr}$ for some h' with $i \le h' \le j-1$. Since $\alpha(q_{h'})\downarrow$ by (10), then there would exist some $\delta \in SEL^+$ such that $\delta\alpha(q_{h'})\downarrow$ and $[\delta\alpha(q_{h'})] \not\le [\alpha(q_{h'})]_\delta = [q_{h'}]_{\delta\circ\alpha}$. But $\delta\circ\alpha \lessdot \alpha$ contradicts the minimality of α. ∇ Thus we have established

(11) $\alpha(q_h) \in T_{\Sigma gr}$, for each h with $i \le h \le j-1$.

Now suppose that $\alpha \ne \alpha_{i-1}$. Then $\alpha \lessdot \alpha_{i-1}$, $\alpha \perp \alpha_{i-1}$ or $a \gtrdot \alpha_{i-1}$, i.e. $\alpha \lessdot \alpha_{i-1}$ or $\alpha \not\lessdot \alpha_{i-1}$:

Case (i) $\alpha \lessdot \alpha_{i-1}$: Since $q_{i-1} \xrightarrow[\alpha_{i-1}]{}_{\mathcal{R}} q_i$ and $\alpha(q_i)\downarrow$ by (10), we can use Lemma 4.5 (2) to infer that $[\alpha(q_i)] \le [q_i]_\alpha$ which contradicts (10). ∇ ∅

Case (ii) $\alpha \not\lessdot \alpha_{i-1}$: Using $q_{i-1} \xrightarrow[\alpha_{i-1}]{}_{\mathcal{R}} q_i$ we obtain

$[\alpha(q_{i-1})]=[\alpha(q_i)]$ and $[q_{i-1}]_\alpha=[q_i]_\alpha$ by Lemma 3.1 (3) and 3.1 (4). Hence by (8) $[\alpha(q_i)]\leq[q_i]_\alpha$ which contradicts (10). ∇ ∅

Hence

(12) $\alpha=\alpha_{i-1}$

and by a similar argument we can prove that

(13) $\alpha=\alpha_{j-1}$.

From (12) and (13) we infer (2), (12) and (11) give us (3) and finally we obtain (4) by (12) and (10). ⊠

<u>Lemma 5.2</u> (Shift-Up Lemma)
Let $\mathcal{R}$ be a Σ-maximal ground term rewriting system and
$q_1 \underset{\alpha_1}{\longrightarrow} q_2 \ldots q_n \underset{\alpha_n}{\longrightarrow} q_{n+1}$, $n\geq 1$, be an $\mathcal{R}$-rewrite.

If $\{\alpha_1(q_1),\ldots,\alpha_1(q_{n+1})\}\subset T_{\Sigma gr}$, then there exist $\beta_1,\ldots,\beta_n\in SEL^*$ such that for each h with $1\leq h\leq n$:

(1) $q_1 \underset{\beta_1}{\longrightarrow}_{\mathcal{R}} q_2 \underset{\beta_2}{\longrightarrow}_{\mathcal{R}} q_3 \quad \ldots \quad q_n \underset{\beta_n}{\longrightarrow}_{\mathcal{R}} q_{n+1}$,

(2) $\beta_h=\alpha_1$, if $\alpha_h \overset{\circ}{<} \alpha_1$, and

(3) $\beta_h=\alpha_h$, if $\alpha_h \overset{\circ}{\not<} \alpha_1$.

Proof The proof is by induction on n.

Base Case n=1: The lemma holds trivially. ∅

Induction Step: Let $q_1 \underset{\alpha_1}{\longrightarrow} q_2 \ldots q_n \underset{\alpha_n}{\longrightarrow} q_{n+1} \underset{\alpha_{n+1}}{\longrightarrow} q_{n+2}$ be an $\mathcal{R}$-rewrite with $\{\alpha_1(q_1),\ldots,\alpha_1(q_{n+1}),\alpha_1(q_{n+2})\}\subset T_{\Sigma gr}$. Our induction hypothesis is to assume that there exist some $\beta_1,\ldots,\beta_n\in SEL^*$

such that conditions (1), (2) and (3) are satisfied for each h with $1 \leq h \leq n$.

For $\alpha_{n+1} \not\leq \alpha_1$ we define $\beta_{n+1} = \alpha_{n+1}$, and with the induction hypothesis $\beta_1, \ldots, \beta_{n+1}$ is a sequence of selectors with the desired properties.

If $\alpha_{n+1} \leq \alpha_1$, then $\alpha_{n+1} = \beta \circ \alpha_1$ for some $\beta \in SEL^+$, i.e. $q_{n+1} \xrightarrow[\beta \circ \alpha_1]{} {}_{\mathcal{R}} q_{n+2}$.

Now by Lemma 3.3 (3) $\alpha_1(q_{n+1}) \underset{\beta}{\overrightarrow{}}_{\mathcal{R}} \alpha_1(q_{n+2})$ and $q_{n+1} \underset{\alpha_1}{\sim} q_{n+2}$.

Since $\alpha_1(q_{n+1}), \alpha_1(q_{n+2}) \in T_{\Sigma gr}$ by assumption, we can write $\alpha_1(q_{n+1}) \rightarrow_{\Sigma\mathcal{R}} \alpha_1(q_{n+2})$ and because $\mathcal{R}$ is Σ-maximal we have $\alpha_1(q_{n+1}) \Rightarrow_{\mathcal{R}} \alpha_1(q_{n+2})$.

With $q_{n+1} \underset{\alpha_1}{\sim} q_{n+2}$ we obtain $q_{n+1} \underset{\alpha_1}{\longrightarrow}_{\mathcal{R}} q_{n+2}$ and setting $\beta_{n+1} = \alpha_1$,

by the induction hypothesis $\beta_1, \ldots, \beta_{n+1}$ is a sequence of selectors such that for each h with $1 \leq h \leq n+1$, conditions (1), (2) and (3) are satisfied. ⌀ ⊠

<u>Lemma 5.3</u> (Shift-Left Lemma)
Let $\mathcal{R}$ be a ground term rewriting system and $q_1 \underset{\alpha_1}{\longrightarrow} q_2 \ldots q_n \underset{\alpha_n}{\longrightarrow} q_{n+1}$, $n \geq 1$, be an $\mathcal{R}$-rewrite. If $\alpha_i \mathrel{\underline{\perp}} \alpha_n$ for each i with $1 \leq i \leq n$, then there exist $r_2, \ldots, r_n \in T_{gr}$ such that

$$q_1 \underset{\alpha_n}{\longrightarrow}_{\mathcal{R}} r_2 \underset{\alpha_1}{\longrightarrow}_{\mathcal{R}} r_3 \ldots r_n \underset{\alpha_{n-1}}{\longrightarrow}_{\mathcal{R}} q_{n+1}.$$

Proof The proof is by induction on n.

Base Case n=1: The lemma holds trivially. ⌀

Induction Step: Let $q_1 \underset{\alpha_1}{\rightarrow} q_2 \ldots q_{n+1} \underset{\alpha_{n+1}}{\longrightarrow} q_{n+2}$ be an $\mathcal{R}$-rewrite such that $\alpha_i \perp \alpha_{n+1}$ for each i with $1 \leq i \leq n+1$.

Our induction hypothesis is to assume that the lemma holds for all $\mathcal{R}$-rewrites with length at most n. Hence we are allowed to assume that $q_2 \underset{\alpha_{n+1}}{\longrightarrow}_{\mathcal{R}} r_3 \underset{\alpha_2}{\rightarrow}_{\mathcal{R}} r_4 \ldots r_{n+1} \underset{\alpha_n}{\rightarrow}_{\mathcal{R}} q_{n+2}$ for some $r_3, \ldots, r_{n+1} \in T_{gr}$. From $q_1 \underset{\alpha_1}{\rightarrow}_{\mathcal{R}} q_2 \underset{\alpha_{n+1}}{\longrightarrow}_{\mathcal{R}} r_3$ and $\alpha_1 \perp \alpha_{n+1}$ we infer by Lemma 3.3 (7) the existence of some $r_2 \in T_{gr}$ such that $q_1 \underset{\alpha_{n+1}}{\longrightarrow}_{\mathcal{R}} r_2 \underset{\alpha_1}{\rightarrow}_{\mathcal{R}} r_3$. Hence we have found some $r_2, r_3, \ldots, r_{n+1} \in T_{gr}$ such that $q_1 \underset{\alpha_{n+1}}{\longrightarrow}_{\mathcal{R}} r_2 \underset{\alpha_1}{\rightarrow}_{\mathcal{R}} r_3 \ldots r_{n+1} \underset{\alpha_n}{\rightarrow}_{\mathcal{R}} q_{n+2}$. ⊠ ⊠

<u>Lemma 5.4</u> (Reduction Lemma)
Let $\mathcal{R}$ be a Σ-maximal ground term rewriting system and $q_1 \underset{\alpha}{\rightarrow} q_2 \ldots q_n \underset{\alpha}{\rightarrow} q_{n+1}$, $n \geq 1$, be an $\mathcal{R}$-rewrite. Then $q_1 \underset{\alpha}{\longrightarrow}_{\mathcal{R}} q_{n+1}$.

Proof For each i with $1 \leq i \leq n$ we know that $q_i \underset{\alpha}{\sim} q_{i+1}$ and $\alpha(q_i) \Rightarrow_{\mathcal{R}} \alpha(q_{i+1})$, hence $\alpha(q_i) \overset{+}{\rightarrow}_{\Sigma\mathcal{R}} \alpha(q_{i+1})$ because $\mathcal{R}$ is Σ-maximal. But then $\alpha(q_1) \overset{+}{\rightarrow}_{\Sigma\mathcal{R}} \alpha(q_{n+1})$ since $\overset{+}{\rightarrow}_{\Sigma\mathcal{R}}$ is transitive and finally $\alpha(q_1) \Rightarrow_{\mathcal{R}} \alpha(q_{n+1})$ by the Σ-maximality of $\mathcal{R}$. Because of $q_i \underset{\alpha}{\sim} q_{i+1}$ and since $\underset{\alpha}{\sim}$ is transitive we have $q_1 \underset{\alpha}{\sim} q_{n+1}$, hence with $\alpha(q_1) \Rightarrow_{\mathcal{R}} \alpha(q_{n+1})$ we infer $q_1 \underset{\alpha}{\longrightarrow}_{\mathcal{R}} q_{n+1}$. ⊠

Finally, we prove the main theorem of this chapter, the Σ-Rewrite Theorem:

Theorem 5.5 (Σ-Rewrite Theorem)

If $\mathcal{R}$ is a Σ-maximal ground term rewriting system, then $\overset{+}{\rightarrow}_{\mathcal{R}} \cap (T_{\Sigma gr} \times T_{\Sigma gr}) = \overset{+}{\rightarrow}_{\Sigma\mathcal{R}}$.

Proof "$\supset$" Obvious, because $\overset{+}{\rightarrow}_{\Sigma\mathcal{R}} \subset \overset{+}{\rightarrow}_{\mathcal{R}}$ and $\overset{+}{\rightarrow}_{\Sigma\mathcal{R}} \subset (T_{\Sigma gr} \times T_{\Sigma gr})$.

"$\subset$" Let us assume by way of contradiction that there exist a pair of Σ-ground terms q_1 and q_{n+1} such that $q_1 \overset{+}{\rightarrow}_{\mathcal{R}} q_{n+1}$ but $q_1 \not\overset{+}{\rightarrow}_{\Sigma\mathcal{R}} q_{n+1}$.

Let $q_1 \underset{\alpha_1}{\longrightarrow} q_2 \ldots q_n \underset{\alpha_n}{\longrightarrow} q_{n+1}$

be an $\mathcal{R}$-rewrite of q_{n+1} from q_1 with minimal length. We know that $n \geq 2$, because $q_1, q_{n+1} \in T_{\Sigma gr}$ and $q_1 \not\overset{+}{\rightarrow}_{\Sigma\mathcal{R}} q_{n+1}$. But then $\{q_2, \ldots, q_n\} \not\subset T_{\Sigma gr}$ and by the Σ-Rewrite Lemma (5.1) there exist indices i and j with $2 \leq i < j \leq n+1$ such that for each h with $i \leq h \leq j-1$

(1) $\alpha_{j-1} = \alpha_{i-1}$,

(2) $\alpha_{i-1}(q_h) \in T_{\Sigma gr}$, and

(3) $[\alpha_{i-1}(q_h)] \not\subseteq [q_h]_{\alpha_{i-1}}$.

Now consider the $\mathcal{R}$-rewrite

(4) $q_{i-1} \underset{\alpha_{i-1}}{\longrightarrow} q_i \ldots q_{j-1} \underset{\alpha_{j-1}}{\longrightarrow} q_j$.

Since $\alpha_{i-1}(q_{i-1}) \Rightarrow_{\mathcal{R}} \alpha_{i-1}(q_i)$ and $\alpha_{j-1}(q_{j-1}) \Rightarrow_{\mathcal{R}} \alpha_{j-1}(q_j)$, we know by Lemma 4.5 (1) that $\alpha_{i-1}(q_{i-1}) \in T_{\Sigma gr}$ and that $\alpha_{j-1}(q_j) \in T_{\Sigma gr}$, because $\mathcal{R}$ is Σ-maximal. Now by (1) and (2) $\{\alpha_{i-1}(q_{i-1}), \ldots, \alpha_{i-1}(q_j)\} \subset T_{\Sigma gr}$ and hence by the Shift-Up Lemma (5.2) there exist $\beta_{i-1}, \ldots, \beta_{j-1} \in SEL^*$ such that for each h with $i-1 \leq h \leq j-1$

(5) $q_{i-1} \underset{\beta_{i-1}}{\longrightarrow}_{\mathcal{R}} q_i \underset{\beta_i}{\rightarrow}_{\mathcal{R}} q_{i+1} \ldots q_{j-1} \underset{\beta_{j-1}}{\longrightarrow}_{\mathcal{R}} q_j$,

(6) $\beta_h=\alpha_{i-1}$, if $\alpha_h \mathring{<} \alpha_{i-1}$, and

(7) $\beta_h=\alpha_h$, if $\alpha_h \not\mathring{<} \alpha_{i-1}$

From (1) and (7) we infer

(8) $\beta_{j-1}=\alpha_{i-1}$.

Now we prove that $\beta_h \underline{\perp} \alpha_{i-1}$ for each h with $i \leq h \leq j-1$:

Case (i) $\alpha_{i-1} \mathring{>} \alpha_h$: Then by (6) $\beta_h=\alpha_{i-1}$, i.e. $\beta_h \underline{\perp} \alpha_{i-1}$ by definition of $\underline{\perp}$. ☒

Case (ii) $\alpha_{i-1} \underline{\perp} \alpha_h$: Then $\alpha_h \not\mathring{<} \alpha_{i-1}$, hence by (7) $\beta_h=\alpha_h$, i.e. $\beta_h \underline{\perp} \alpha_{i-1}$. ☒

Case (iii) $\alpha_{i-1} \mathring{<} \alpha_h$: Using $q_h \xrightarrow[\alpha_h]{}_{\mathcal{R}} q_{h+1}$ and $\alpha_{i-1}(q_h)\downarrow$ by (2), we obtain by Lemma 4.5 (2) that $[\alpha_{i-1}(q_h)] \leq [q_h]_{\alpha_{i-1}}$, i.e. a contradiction to (3). ∇ ☒

Hence, using (7) and (8), we can write

(9) $\beta_h \underline{\perp} \beta_{j-1}$, for each h with $i-1 \leq h \leq j-1$.

Now with (5) and (9) we can use the Shift-Left Lemma (5.3) to infer the existence of some $r_i, \ldots, r_{j-1} \in T_{gr}$ such that

(10) $q_{i-1} \xrightarrow[\beta_{j-1}]{}_{\mathcal{R}} r_i \xrightarrow[\beta_{i-1}]{}_{\mathcal{R}} r_{i+1} \cdots r_{j-1} \xrightarrow[\beta_{j-2}]{}_{\mathcal{R}} q_j$

and in particular

(11) $q_{i-1} \xrightarrow[\alpha_{i-1}]{}_{\mathcal{R}} r_i \xrightarrow[\alpha_{i-1}]{}_{\mathcal{R}} r_{i+1}$

because $\beta_{j-1}=\alpha_{i-1}$ by (8) and $\beta_{i-1}=\alpha_{i-1}$ by (7). But with (11) we can use the Reduction Lemma (5.4) to obtain

(12) $q_{i-1} \xrightarrow[\alpha_{i-1}]{}_{\mathcal{R}} r_{i+1}$.

Summarizing, we have found an $\mathcal{R}$-rewrite

$$q_1 \xrightarrow[\alpha_1]{} q_2 \ldots q_{i-1} \xrightarrow[\alpha_{i-1}]{} r_{i+1} \xrightarrow[\beta_i]{} r_{i+2} \ldots r_{j-1} \xrightarrow[\beta_{j-2}]{} q_j \xrightarrow[\alpha_j]{} q_{j+1} \ldots q_n \xrightarrow[\alpha_n]{} q_{n+1}$$

of q_{n+1} from q_1 with length n-1, i.e. the $\mathcal{R}$-rewrite

$$q_1 \xrightarrow[\alpha_1]{} q_2 \ldots q_n \xrightarrow[\alpha_n]{} q_{n+1}$$

initially given was not of minimal length. ∇ ⊠

Note that the Σ-Rewrite Theorem obviously also holds for $\mathcal{R}$-rewrites of ground literals, i.e.

$$\overset{+}{\rightarrow}_{\mathcal{R}} \cap (LIT_{\Sigma gr} \times LIT_{\Sigma gr}) = \overset{+}{\rightarrow}_{\Sigma\mathcal{R}}.$$

6 Completeness of the ΣRP-Calculus – The Ground Case

In this chapter we will prove the ground completeness theorem for the ΣRP-calculus, i.e. we will show that $S^E_{\Sigma gr} \vdash_\Sigma \square$, if $S_{\Sigma gr}$ is E-unsatisfiable.

Let us outline the proof idea before we go into formal details. The differences between Σ-deductions and ordinary deductions are

- the restriction to well-sorted substitutions in forming Σ-resolvents, Σ-factors and Σ-paramodulants, and
- the application of a coercion substitution on forming a Σ-paramodulant.

Now with ε being the only unifier in the ground case, and of course $\varepsilon \in SUB_\Sigma$, we know that each deduction $\vdash_R$ from a well-sorted ground clause set is, obviously, also a Σ-deduction $\vdash_{\Sigma R}$, i.e. for all well-sorted clause sets S we have:

$$\{C \in \mathcal{L}_{\Sigma gr} \mid S_{\Sigma gr} \vdash_{\Sigma R} C\} = \{C \in \mathcal{L}_{gr} \mid S_{\Sigma gr} \vdash_R C\}.$$

Hence each refutation of a well-sorted ground clause set is always a Σ-refutation and therefore the ΣRP-calculus without paramodulation is complete in the ground case.

However, this result does not hold for paramodulation. The reason is that the application of coercion substitutions is also effective in the ground case: with $\varepsilon \in SUB_\Sigma$ being the only unifier in the ground case, we have that $Par_\Sigma(C,L,D,E(qr),\alpha,\nu \circ \varepsilon) = \nu(C-L) \cup \nu(D-E(qr)) \cup \{\nu K\}$ is a Σ-paramodulant of the Σgr-clauses C and D upon $L \in C$ and $E(qr) \in D$ iff

(1) $\alpha(L) = t_1$,

(2) $L \underset{\alpha}{\sim} K$ and $\alpha(K)=t_2$, and

(3) $\nu \in mc\mathcal{U}_\Sigma(\{z,t_2\})$,

where $t_1=q$, $t_2=r$ or $t_1=r$, $t_2=q$ and $z \in \mathcal{V}_{[L]\alpha} \setminus \mathcal{V}(C \cup D)$. Since $t_2 \in T_{gr}$, $\nu=\{z \leftarrow t_2\}$ is the only unifier of $\{z,t_2\}$ (and of course ν is an mgu of $\{z,t_2\}$), and therefore $\nu \in mc\mathcal{U}_\Sigma(\{z,t_2\})$ iff $\nu \in SUB_\Sigma$. But $\nu \in SUB_\Sigma$ iff $[t_2] \leq [z] = [L]_\alpha$ (cf. Lemma 7.1) and using (2) we obtain with Lemma 4.2 $K \in LIT_{\Sigma gr}$. Hence for the Σ-ground case we can replace the above condition (3) by the equivalent condition (3') $K \in LIT_{\Sigma gr}$. Since $z \notin \mathcal{V}(C \cup D)$ we have $\nu_{|\mathcal{V}(C \cup D)} = \varepsilon$ and we obtain finally

$$\begin{aligned} Par_\Sigma(C,L,D,E(qr),\alpha,\nu \circ \varepsilon) &= \nu(C-L) \cup \nu(D-E(qr)) \cup \{\nu K\} \\ &= (C-L) \cup (D-E(qr)) \cup \{K\} \\ &= Par(C,L,D,E(qr),\alpha,\varepsilon). \end{aligned}$$

Hence Σ-paramodulation differs from paramodulation in the Σ-ground case only in the restriction that modulant literals have to be well-sorted. Therefore we can write

$$\{C \in \mathcal{L}_{\Sigma gr} \mid S_{\Sigma gr} \vdash_{\Sigma P} C\} \subset \{C \in \mathcal{L}_{gr} \mid S_{\Sigma gr} \vdash_P C\}.$$

So Σ-paramodulation can be considered as a real refinement of paramodulation even in the ground case. Obviously the completeness of the ΣRP-calculus in the ground case is equivalent to the completeness of this refinement, which can be formulated as

(*)
if $\{C \in \mathcal{L}_{gr} \mid S^E_{\Sigma gr} \vdash_P C\}$ is unsatisfiable,
then $\{C \in \mathcal{L}_{\Sigma gr} \mid S^E_{\Sigma gr} \vdash_{\Sigma P} C\}$ is unsatisfiable.

To prove (*) we introduce the following clause sets:

Definition 6.1

$Par(S) = \{C \in \mathcal{L}_{gr} \mid S^{E}_{\Sigma gr} \vdash_{P} C\}$, $Par_{\Sigma}(S) = \{C \in \mathcal{L}_{\Sigma gr} \mid S^{E}_{\Sigma gr} \vdash_{\Sigma P} C\}$, and

$RPar(S) = \{C \in \mathcal{L}_{gr} \mid S^{E}_{\Sigma gr} \vdash_{P} C$, such that no clause in $\vdash_{P}$ is obtained by paramodulating into a positive equality literal$\}$. ⊠

Since the unsatisfiability of Par(S) is equivalent to the unsatisfiability of RPar(S), cf. [Loveland 1978], (*) is proved if we have shown that the satisfiability of $Par_{\Sigma}(S)$ implies the satisfiability of RPar(S). This is achieved in the following way:

First we introduce the notion of a ΣE-restricted interpretation and we show that $Par_{\Sigma}(S)$ possesses a model, which is a ΣE-restricted interpretation, whenever $Par_{\Sigma}(S)$ is satisfiable. Next we prove that each ΣE-restricted interpretation contains a Σ-maximal and symmetric term rewriting system. This fact is used to introduce the rewrite-closure I* of a ΣE-restricted interpretation I and to prove that I* is again an interpretation. Moreover, we show that the rewrite-closure I* of I is a model of RPar(S), provided I satisfies $Par_{\Sigma}(S)$.

The principal idea in the completeness proof for paramodulation is to show that each satisfiable ground clause set which is closed under paramodulation possesses an E-model: it is proved that each minimal model of Par(S) is an E-model [Wos and Robinson 1973].

However, this proof idea does not work for Σ-paramodulation directly because a minimal model M of $Par_{\Sigma}(S)$ is not an E-model but only contained in an E-model M* of $Par_{\Sigma}(S)$ (if M* exists at all). The reason is that all literals in $Par_{\Sigma}(S)$ are well-sorted and therefore $M = M^* \cap LIT_{\Sigma gr}$.

To deal formally with interpretations, which correspond to an E-interpretation on $LIT_{\Sigma gr}$, we introduce the notion of a ΣE-restricted interpretation:

Definition 6.2

An interpretation I is called Σ-reflexive iff $E(qq)\in I$ for each $q\in T_{\Sigma gr}$. We say that I is ΣE-closed iff for each $L\in I$ and for each $K\in LIT_{\Sigma gr}$, $K\in I$ whenever $L\rightarrow_{\mathcal{R}(I)}K$. I is called a ΣE-restricted interpretation iff I is Σ-reflexive, I is ΣE-closed and $I\subset LIT_{\Sigma gr}$. ⊠

We prove that each satisfiable $Par_\Sigma(S)$ has a ΣE-restricted model:

Lemma 6.1

If $Par_\Sigma(S)$ is satisfiable, then it possesses a model, which is a ΣE-restricted interpretation.

Proof Let M be a minimal model of $Par_\Sigma(S)$. We show that M is a ΣE-restricted interpretation:

M is Σ-reflexive:

Obvious, because $\{E(qq)\}\in S^E_{\Sigma gr}\subset Par_\Sigma(S)$ for each $q\in T_{\Sigma gr}$ and M satisfies $Par_\Sigma(S)$. ∅

M is ΣE-closed:

Let $L\in M$ and $K\in LIT_{\Sigma gr}$ such that $L\rightarrow_{\mathcal{R}(M)}K$, i.e. $L\rightarrow_{\alpha M}K$ for some $\alpha\in SEL^+$. Now assume by way of contradiction, that $M\cap C\neq\{L\}$ for each $C\in Par_\Sigma(S)$. With $M\cap C\neq\emptyset$ for each $C\in Par_\Sigma(S)$ we obtain that $(M-L)\cap C\neq\emptyset$, i.e. M-L satisfies $Par_\Sigma(S)$ and therefore M is not minimal. ∇ Hence

(1) $M\cap C_L=\{L\}$, for some $C_L\in Par_\Sigma(S)$,

and by an analogous argument

(2) $M \cap C_E = \{E(\alpha(L)\alpha(K))\}$, for some $C_E \in Par_\Sigma(S)$.

Let C be the paramodulant of C_L and C_E upon L and $E(\alpha(L)\alpha(K))$, i.e.

(3) $C=(C_L-L) \cup (C_E-E(\alpha(L)\alpha(K))) \cup \{K\}$.

We know that $[\alpha(K)]\leq[K]_\alpha$ because $K\in LIT_{\Sigma gr}$, and that $[K]_\alpha=[L]_\alpha$ by Lemma 3.1(4), i.e. $[\alpha(K)]\leq[L]_\alpha$. Hence C is a Σ-paramodulant, i.e. $C\in Par_\Sigma(S)$ and therefore

(4) $M\cap C\neq\emptyset$.

Using (1), (2) and (3) we infer that $M\cap C=M\cap\{K\}$, hence by (4) $K\in M$, thus M is ΣE-closed. ∅

$M\subset LIT_{\Sigma gr}$:
Suppose that $L\in M$ for some $L\notin LIT_{\Sigma gr}$. Then $L\notin M\cap C$ for each $C\in Par_\Sigma(S)$, because each clause in $Par_\Sigma(S)$ contains only Σ-literals. Hence $M-L$ is also a model of $Par_\Sigma(S)$, i.e. M is not minimal. ∇ ∅ ⊠

We shall extend the ΣE-restricted model of $Par_\Sigma(S)$, which must exist by Lemma 6.1, to a model of RPar(S). To do so, we use the following characteristic property of ΣE-restricted interpretations:

<u>Lemma 6.2</u>
If I is a ΣE-restricted interpretation, then $\mathcal{R}(I)$ is a Σ-maximal and symmetric term rewriting system.

Proof First we prove that $\mathcal{R}(I)$ is Σ-maximal, i.e. $\Rightarrow_{\mathcal{R}(I)} = \overset{+}{\rightarrow}_{\Sigma\mathcal{R}(I)}$:

"$\subset$": Obviously $\Rightarrow_{\mathcal{R}(I)}\subset\rightarrow_{\mathcal{R}(I)}$ and $\Rightarrow_{\mathcal{R}(I)}\subset(T_{\Sigma gr}\times T_{\Sigma gr})$ because $I\subset LIT_{\Sigma gr}$. Hence $\Rightarrow_{\mathcal{R}(I)}\subset\rightarrow_{\Sigma\mathcal{R}(I)}\subset\overset{+}{\rightarrow}_{\Sigma\mathcal{R}(I)}$. ∅

"⊃": Let $q_1, q_{n+1} \in T_{\Sigma gr}$ such that $q_1 \overset{+}{\to}_{\Sigma \mathcal{R}(I)} q_{n+1}$, i.e. there exist $q_2, \ldots, q_n \in T_{\Sigma gr}$, $n \geq 1$, and some $\alpha_1, \ldots, \alpha_n \in SEL^*$ such that

$$(1) \quad q_1 \underset{\alpha_1}{\to} q_2 \ldots q_n \underset{\alpha_n}{\to} q_{n+1}$$

is an $\mathcal{R}(I)$-rewrite. We prove by induction on the length n of the $\mathcal{R}(I)$-rewrite (1), that $q_1 \Rightarrow_{\mathcal{R}(I)} q_{n+1}$.

Base Case n=1:

Let $\alpha \in SEL^+$ such that $\alpha(E(t_1 t_2)) = t_2$ for all $t_1, t_2 \in T$. Then $\alpha(E(q_1 q_1)) = q_1 \underset{\alpha_1}{\to}_{\mathcal{R}(I)} q_2 = \alpha(E(q_1 q_2))$ and with $E(q_1 q_1) \underset{\alpha}{\sim} E(q_1 q_2)$ we infer by Lemma 2.2 (3) that

$$(2) \quad E(q_1 q_1) \xrightarrow[\alpha_1 \circ \alpha]{} {}_{\mathcal{R}(I)} E(q_1 q_2).$$

Since $q_1, q_2 \in T_{\Sigma gr}$ we know that $E(q_1 q_2) \in LIT_{\Sigma gr}$. By the Σ-reflexivity of I we obtain $E(q_1 q_1) \in I$ and with (2) $E(q_1 q_2) \in I$, because I is ΣE-closed, i.e. $q_1 \Rightarrow_{\mathcal{R}(I)} q_2$. ∅

Induction Step:

Our induction hypothesis is to assume that $q \Rightarrow_{\mathcal{R}(I)} r$ for each $\mathcal{R}(I)$-rewrite of r from q with length at most n, provided $q \overset{+}{\to}_{\Sigma \mathcal{R}(I)} r$. Let

$$q_1 \underset{\alpha_1}{\to} q_2 \ldots q_{n+1} \underset{\alpha_{n+1}}{\longrightarrow} q_{n+2}$$

be an $\mathcal{R}(I)$-rewrite such that $q_1, \ldots, q_{n+2} \in T_{\Sigma gr}$, i.e.

$$q_1 \overset{+}{\to}_{\Sigma \mathcal{R}(I)} q_{n+2}.$$

By the induction hypothesis we obtain that $q_1 \Rightarrow_{\mathcal{R}(I)} q_2$, i.e. $E(q_1 q_2) \in I$ and also $q_2 \Rightarrow_{\mathcal{R}(I)} q_{n+2}$, i.e. $E(q_2 q_{n+2}) \in I$. Since q_1, $q_{n+2} \in T_{\Sigma gr}$, we know that $E(q_1 q_{n+2}) \in LIT_{\Sigma gr}$ and choosing $\alpha \in SEL^+$ as in the base case we obtain $E(q_1 q_2) \underset{\alpha}{\sim} E(q_1 q_{n+2})$ and $E(\alpha(E(q_1 q_2))$ $\alpha(E(q_1 q_{n+2}))) = E(q_2 q_{n+2}) \in I$, i.e.

(3) $E(q_1q_2) \vec{_{\alpha I}} E(q_1q_{n+2})$.

With $E(q_1q_2) \in I$ and $E(q_1q_{n+2}) \in LIT_{\Sigma gr}$ we obtain with (3) and the ΣE-closure of I, $E(q_1q_{n+2}) \in I$, i.e. $q_1 \Rightarrow_{\mathcal{R}(I)} q_{n+2}$. ∅

To prove that $\mathcal{R}(I)$ is symmetric, suppose that $q \Rightarrow_{\mathcal{R}(I)} r$ for any $q,r \in T_{\Sigma gr}$. Let $\alpha \in SEL^+$ such that $\alpha(E(t_1t_2))=t_1$ for all $t_1,t_2 \in T$. Then $E(qq) \underset{\alpha}{\sim} E(rq)$ and $E(\alpha(qq)\alpha(E(rq))) = E(qr) \in I$, i.e.

$$E(qq) \vec{_{\alpha I}} E(rq).$$

Obviously $E(rq) \in LIT_{\Sigma gr}$ and by the Σ-reflexivity of I, $E(qq) \in I$. Hence $E(rq) \in I$, because I is ΣE-closed, i.e. $r \Rightarrow_{\mathcal{R}(I)} q$. ∅ ⊠

We construct the so called rewrite-closure I* of a ΣE-restricted interpretation I by extending I with all the ground literals which are obtained by $\overset{+}{\to}_{\mathcal{R}(I)}$ from all non-equality atoms in I:

<u>Definition 6.3</u>

For a ΣE-restricted interpretation I the rewrite-closure I* of I is given by

$$I^* = I \cup \{K \in LIT_{gr} \setminus AT^E_{gr} \mid L \overset{+}{\to}_{\mathcal{R}(I)} K \text{ for some } L \in I\}. \text{ ⊠}$$

Obviously the rewrite-closure I* may also contain ill-sorted ground literals. But surprisingly we do not introduce new well-sorted ground literals when we extend I to I*:

<u>Lemma 6.3</u>

If I is a ΣE-restricted interpretation, then $I = I^* \cap LIT_{\Sigma gr}$.

Proof "⊂" Obvious, because $I \subset I^*$ and $I \subset LIT_{\Sigma gr}$. ∅

"$\supset$" Let $K \in I^* \cap LIT_{\Sigma gr}$. If $K \in I$ the case is proved. For $K \in I^* \setminus I$, there exists some $L \in I$ such that $L \overset{+}{\rightarrow}_{\mathcal{R}(I)} K$. By assumption $K \in LIT_{\Sigma gr}$ and with $I \subset LIT_{\Sigma gr}$ we know that $L \in LIT_{\Sigma gr}$. By Lemma 6.2 we find that $\mathcal{R}(I)$ is Σ-maximal, hence by the Σ-Rewrite Theorem 5.5 $L \overset{+}{\rightarrow}_{\Sigma\mathcal{R}(I)} K$, i.e. there exists an $\mathcal{R}(I)$-rewrite

$$L = L_1 \underset{\alpha_1}{\longrightarrow} L_2 \; \ldots \; L_n \underset{\alpha_n}{\longrightarrow} L_{n+1} = K, \text{ such that } \{L_1, \ldots, L_{n+1}\} \subset LIT_{\Sigma gr}.$$

It is easily verified by induction on n using the ΣE-closure of I, that $\{L_1, \ldots, L_{n+1}\} \subset I$, hence $K \in I$. ⊠

Now we prove that the rewrite closure of a ΣE-restricted interpretation is always an interpretation:

<u>Lemma 6.4</u>
If I is a ΣE-restricted interpretation, then I^* is an interpretation.

Proof Assume by contradiction that I^* is not an interpretation, i.e. $\{Q, Q^c\} \subset I^*$ for some ground literal Q. Then by Definition 6.3 there exist literals L and K^c in I such that

$$L \overset{*}{\rightarrow}_{\mathcal{R}(I)} Q \text{ and } K^c \overset{*}{\rightarrow}_{\mathcal{R}(I)} Q^c.$$

With $K^c \overset{*}{\rightarrow}_{\mathcal{R}(I)} Q^c$ we obtain $K \overset{*}{\rightarrow}_{\mathcal{R}(I)} Q$ and since $\mathcal{R}(I)$ is symmetric by Lemma 6.2 we have $Q \overset{*}{\rightarrow}_{\mathcal{R}(I)} K$. Finally with $L \overset{*}{\rightarrow}_{\mathcal{R}(I)} Q$ we find by the transitivity of $\overset{*}{\rightarrow}_{\mathcal{R}(I)}$ that $L \overset{*}{\rightarrow}_{\mathcal{R}(I)} K$.

Now suppose that $K \in AT^E_{gr}$. Then $Q \in AT^E_{gr}$ and by Definition 6.3 $Q \in I$. But then $Q \in LIT_{\Sigma gr}$, i.e. $Q^c \in LIT_{\Sigma gr}$, hence by Lemma 6.3 $Q^c \in I$ contradicting that I is an interpretation. ∇

So let us assume that $K \notin AT^E_{gr}$. Then by Definition 6.3 $K \in I^*$ and since $K^c \in I$, $K^c \in LIT_{\Sigma gr}$, i.e. $K \in LIT_{\Sigma gr}$. Hence by Lemma 6.3 $K \in I$ and since $K^c \in I$, again I cannot be an interpretation. ∇ ⊠

Using Lemma 6.4 we can construct a model of RPar(S) from a model of $Par_\Sigma(S)$:

Lemma 6.5

If $Par_\Sigma(S)$ is satisfiable, then RPar(S) is satisfiable.

Proof By Lemma 6.1 there exists a model M of $Par_\Sigma(S)$, which is a ΣE-restricted interpretation. Hence by Lemma 6.4, M* is an interpretation. We prove by induction on the length n of a deduction $S^E_{\Sigma gr} \vdash_P C$, that M* satisfies each clause $C \in RPar(S)$.

Base Case n=0: Then $C \in S^E_{\Sigma gr} \subset Par_\Sigma(S)$ and $M^* \cap C \neq \emptyset$, because M satisfies $Par_\Sigma(S)$ and $M \subset M^*$. ☐

Induction Step: Let $C=(C_L-L)\cup(C_E-E(\alpha(L)\alpha(K)))\cup\{K\}$ be a paramodulant of the clauses $C_L, C_E \in RPar(S)$ upon L and $E(\alpha(L)\alpha(K))$. Then

$$L \underset{\alpha}{\sim} K \text{ and } L \notin AT^E_{gr}$$

by Definition 6.1. Our induction hypothesis is to assume that $M^* \cap C_L \neq \emptyset$ and $M^* \cap C_E \neq \emptyset$.

If $M^* \cap (C_L-L) \neq \emptyset$ or $M^* \cap (C_E-E(\alpha(L)\alpha(K))) \neq \emptyset$ then $M^* \cap C \neq \emptyset$ and the induction step is proved. So let us assume that $M^* \cap (C_L-L)=\emptyset$ and $M^* \cap (C_E-E(\alpha(L)\alpha(K)))=\emptyset$. Then by the induction hypothesis $L \in M^*$ and $E(\alpha(L)\alpha(K)) \in M^*$ and by Definition 6.3 $E(\alpha(L)\alpha(K)) \in M$. Using $L \underset{\alpha}{\sim} K$ we obtain $L \underset{\alpha M}{\rightarrow} K$, i.e. $L \overset{*}{\rightarrow}_{\mathcal{R}(M)} K$.

With $L \in M^*$ we find some $Q \in M$ such that $Q \overset{*}{\rightarrow}_{\mathcal{R}(M)} L$, hence $Q \overset{*}{\rightarrow}_{\mathcal{R}(M)} K$. From $L \notin AT^E_{gr}$ we obtain $K \notin AT^E_{gr}$. Hence by Definition 6.3 $K \in M^*$, i.e. $M^* \cap C=\{K\} \neq \emptyset$. ☐ ☒

Using Lemma 6.5 we can prove the main result of this chapter:

Theorem 6.6 (Ground Completeness Theorem for ΣRP)
If $S_{\Sigma gr}$ is E-unsatisfiable, then $S^E_{\Sigma gr} \vdash_\Sigma \Box$.

Proof If $S_{\Sigma gr}$ is E-unsatisfiable, then Par(S) is E-unsatisfiable because $S_{\Sigma gr} \subset Par(S)$. By Theorem 1 from [Wos and Robinson 1973] we infer that Par(S) is unsatisfiable, hence RPar(S) is unsatisfiable [Loveland 1978]. But then by contraposition of Lemma 6.5 $Par_\Sigma(S)$ is unsatisfiable and there exists a finite and unsatisfiable subset P of $Par_\Sigma(S)$ by the Compactness Theorem [Loveland 1978]. Hence $P \vdash_R \Box$ by the completeness of the RP-calculus [Robinson 1965] and since there is no difference between a Σ-deduction $\vdash_{\Sigma R}$ and a deduction $\vdash_R$ from a Σ-ground clause set, we can write $P \vdash_{\Sigma R} \Box$. From $P \subset Par_\Sigma(S)$ we infer that $S^E_{\Sigma gr} \vdash_{\Sigma P} C$ for each $C \in P$ and thus we obtain $S^E_{\Sigma gr} \vdash_\Sigma \Box$. ⊠

7 Many-Sorted Unification

An important result of first-order unification theory is the Unification Theorem [Robinson 1965] which states the existence of a most general unifier for a set of unifiable terms. The Unification Theorem is also a necessary prerequisite to prove the completeness of the RP-calculus in the general case [Robinson 1965, Wos and Robinson 1973].

But unfortunately for many-sorted unification, the Unification Theorem only holds for signatures, where $(\mathcal{S}, <_{\mathcal{S}})$ is a forest structure, i.e. the notion of a most general unifier is too strong to provide us with all the necessary concepts when unification is generalized to the many-sorted case (with $<_{\mathcal{S}} \neq \emptyset$). The situation is similar (but not identical) to that of unification under equational theories, where the notion of an mgu is replaced by the concept of complete and minimal sets of unifiers.

In this chapter we will prove the Σ-Unification Theorem, which states the existence of a finite, complete and minimal set of Σ-unifiers for each Σ-unifiable set of Σ-terms.

The following examples should serve to demonstrate the problems associated with Σ-unification and also to illustrate the results presented subsequently. We start with a lemma, which distinguishes Σ-substitutions from ordinary substitutions:

Lemma 7.1

For each $\sigma \in SUB$, $\sigma \in SUB_\Sigma$ iff $\sigma x \in T_\Sigma$ and $[\sigma x] \leq [x]$ for each $x \in \mathcal{V}$.

Proof '→' Since $\sigma \in SUB_\Sigma$ iff $\sigma T_\Sigma \subset T_\Sigma$, we have $\sigma \mathcal{V} \subset T_\Sigma$, i.e. $\sigma x \in T_\Sigma$ for each $x \in \mathcal{V}$. Now assume by way of contradiction that $[\sigma x'] \nleq [x']$ for some $x' \in \mathcal{V}$. If $f \in \mathcal{F}_{[x'],s}$ for some $s \in \mathcal{S}$, then $f(x') \in T_\Sigma$ but $\sigma f(x') = f(\sigma x') \notin T_\Sigma$ by Lemma 4.3 (2) because $[\sigma x'] \nleq [x'] = [f]_1$. Hence $\sigma T_\Sigma \not\subset T_\Sigma$, i.e. $\sigma \notin SUB_\Sigma$. ∇ ⊠

'←' We prove by structural induction on t that $\sigma t \in T_\Sigma$ for each $t \in T_\Sigma$:

Base Case $t \in \mathcal{C}$: Then $\sigma t = t \in T_\Sigma$ by Lemma 4.3 (1). ⊠

Base Case $t \in \mathcal{V}$: Then $\sigma t \in T_\Sigma$ by assumption. ⊠

Induction Step: Suppose that $t = f(t_1 \ldots t_n) \in T_\Sigma$ and $\sigma t_i \in T_\Sigma$ for each i with $1 \leq i \leq n$. If $t_i \in \mathcal{V}$, then $[\sigma t_i] \leq [t_i]$ by assumption, and if $t_i \notin \mathcal{V}$, then $[\sigma t_i] = [t_i]$ by Lemma 4.1 (2). Hence $[\sigma t_i] \leq [t_i] \leq [f]_i$ for each i, i.e. $\sigma t \in T_\Sigma$ by Lemma 4.3 (2). ⊠ ⊠

Corollary 7.2

If $\sigma \in SUB_\Sigma$ and $t \in T$, then $[\sigma t] \leq [t]$.

Proof If $t \in \mathcal{V}$, then $[\sigma t] \leq [t]$ by Lemma 7.1 and if $t \notin \mathcal{V}$, then $[\sigma t] = [t]$ by Lemma 4.1 (2). ⊠

Example 7.1

Let $(\mathcal{S}, <_\mathcal{S}) = (\mathbb{N}, <_\mathbb{N})$ and $D = \{x, y\}$ with $x \in \mathcal{V}_n$ and $y \in \mathcal{V}_m$, where $\mathbb{N}$ denotes the set of natural numbers ordered by $<_\mathbb{N}$. Then $\sigma_1 = \{x \leftarrow y\}$ and $\sigma_2 = \{y \leftarrow x\}$ are the only mgu's of D. For $n \geq_\mathbb{N} m$, $\sigma_1 \in SUB_\Sigma$, and for $m \geq_\mathbb{N} n$, $\sigma_2 \in SUB_\Sigma$. Hence D possesses a Σ-mgu in both cases. ⊠

Later we will prove that each Σ-unifiable Σ-termset has a Σ-mgu, provided $(\mathcal{S}, <_\mathcal{S})$ is a forest. The reason is that in a forest-structured sort hierarchy each Σ-unifiable pair of variables, as x,y above, can be arranged in an mgu, i.e. $x \leftarrow y$ or $y \leftarrow x$, so that the resulting substitution is well-sorted.

We will also prove that being a forest structure is also a necessary condition for the Robinson Unification Theorem to hold true for many-sorted unification. As a consequence we have to use auxiliary variables in order to obtain complete and minimal sets of Σ-unifiers if non forest-structured sort hierarchies are used.

<u>Example 7.2</u>

Let $\mathcal{S}=S\cup S_k$ with $S=\{a,b\}$ and $S_k=\{n\in\mathbb{N} \mid n<_{\mathbb{N}}k\}$ where $k\in\mathbb{N}^+$ and let $<_{\mathcal{S}}=S_k\times S$. Figure 7.1 is a diagram of the sort hierarchy under consideration, which is obviously not a forest structure:

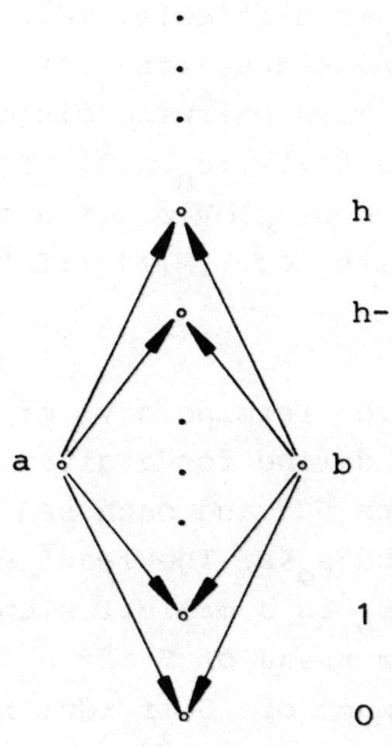

Figure 7.1 The sort hierarchy of Example 7.2

Furthermore let $D=\{x,y\}$ with $x\in\mathcal{V}_a$ and $y\in\mathcal{V}_b$, let $\sigma_1=\{x\leftarrow y\}$, $\sigma_2=\{y\leftarrow x\}$ and let $\theta_h=\{x\leftarrow z_h, y\leftarrow z_h\}$ where $z_h\in\mathcal{V}_h$ for any $h\in S_k$. Since neither $a\not<_{\mathcal{S}}b$ nor $b\not<_{\mathcal{S}}a$, neither σ_1 nor σ_2 are Σ-substitutions, i.e. D has no Σ-mgu. Now consider $U_k=\{\theta_h\in SUB \mid h\in S_k\}$. Obviously $U_k\subset U_\Sigma(D)$ and we can prove that U_k is complete and minimal. ⊠

For the formal treatment of many-sorted unification we need the following additional definitions:

Each unifiable subset D of T induces an equivalence relation $\underset{D}{\sim}$ on $T\times T$ by $q\underset{D}{\sim}r$ iff $\sigma q=\sigma r$ for some mgu σ of D. By the most generality of σ, this definition is independent of σ, since $\tau q=\tau r$ for each mgu τ of D iff $\sigma q=\sigma r$ for some mgu σ of D. As usual $T/\underset{D}{\sim}$ denotes the quotient set of T modulo $\underset{D}{\sim}$. Obviously $U(D)\subset U(M)$ and (because M is unifiable) $|M|<\infty$ for each equivalence class $M\in T/\underset{D}{\sim}$. For $\sigma\in SUB$, a separation of σ is a sequence $\sigma_1,\ldots,\sigma_n$ of substitutions with pairwise disjoint domains and codomains satisfying (1) $\sigma=\sigma_1\circ\ldots\circ\sigma_n$, (2) $COD(\sigma_i)=\{y_i\}\subset\mathcal{V}$ for each i with $1\leq i<n$, and (3) $COD(\sigma_n)\cap\mathcal{V}=\emptyset$. If σ is an mgu of some $D\subset T$, then for all $M\in T/\underset{D}{\sim}$ with $M\subset\mathcal{V}$: $|M|>1$ iff $M=DOM(\sigma_i)\cup COD(\sigma_i)$ for some $i\in\{1,\ldots,n-1\}$.

Following the standard terminology of lattice theory, e.g. [Birkhoff 1967], we define for a given sort hierarchy $(\mathcal{S},<_{\mathcal{S}})$ and for each $s_o\in\mathcal{S}$, each $S\subset\mathcal{S}$ and each $s\in S$: s_o is the least element of S iff $s_o\in S$ and $s_o\leq s$. The least element of S is unique if it exists at all. s_o is a maximal element of S iff $s_o\in S$ and $s_o\not< s$. s_o is a lower bound of S iff $s_o\leq s$. s_o is the greatest lower bound or infimum of S if additionally $s'\leq s_o$ for each lower bound s' of S.

The infimum of S is unique if it exists and will be denoted $\sqcap S$. The least element of S is the infimum of S and if S has an infimum such that $\sqcap S\in S$, then $\sqcap S$ is the least element of S.

The set of all maximal elements in S will be denoted $max_{\mathcal{S}}(S)$, and $lbs_{\mathcal{S}}(S)$ is the set of all lower bounds of S in $\mathcal{S}$, i.e. $max_{\mathcal{S}}(S)=\{s_o \in S \mid s_o \nless_{\mathcal{S}} s \ \forall s \in S\}$ and $lbs_{\mathcal{S}}(S)=\{s_o \in \mathcal{S} \mid s_o \leq_{\mathcal{S}} s \ \forall s \in S\}$. Note that $\sqcap S \in S$ iff $lbs_{\mathcal{S}}(S) \cap S \neq \emptyset$. S is a **chain** in $\mathcal{S}$ iff S is **totally ordered** by $<$, i.e. $s_1 \leq s_2$ or $s_1 \geq s_2$ for all $s_1, s_2 \in S$. We note that any finite chain in $\mathcal{S}$ has a least element. $(\mathcal{S}, <_{\mathcal{S}})$ is called a **meet-semilattice** iff each pair of sort symbols s_1 and s_2 has an infimum.

For $D \subset T$, $[D]$ is the **sortal image** of D defined as $\{[t] \in \mathcal{S} \mid t \in D\}$. Since the sort of a term is determined only by the outermost symbol of t, we have $[\sigma t]=[t]$ and $[\sigma D]=[D]$ for each $\sigma \in SUB$, $t \in T \setminus \mathcal{V}$ and $D \subset T \setminus \mathcal{V}$, cf. Lemma 4.1 (2).

For the remainder of this chapter, we let D be any finite and Σ-unifiable subset of T_Σ for which we are seeking a complete and minimal set of Σ-unifiers. As we have seen in Example 7.2, auxiliary variables can become indispensable in order to obtain such a set. These new variables are formally introduced by certain Σ-substitutions, called **weakening substitutions**, which replace certain variables in the set D. Similar to coercion substitutions, cf. Chapter 4, the application of a weakening substitution can be thought of as a kind of **coercion process** which is necessary in programming languages with **types**, e.g. PASCAL or ADA, whenever the types of a variable and an expression do not agree in an assignment.

Since there are several distinct variable replacements we have also several distinct weakening substitutions, which are collected in a so-called **most general weakening set** W(D) of D.

Let us look closer at the introduction of new variables before we go into formal details. Suppose that during the computation of a complete and minimal set of Σ-unifiers of D, as a subproblem some variables $\{x_1,...,x_n\}=V \subset \mathcal{V}(D)$ have to be Σ-unified 'as generally as possible'.

If $\sqcap[V]\in[V]$, then $\sqcap[V]=[x]$ for some $x\in V$, and $\lambda_x=\{x_1\leftarrow x,\ldots,x_n\leftarrow x\}$ is obviously a Σ-mgu of V, i.e. we do not need auxiliary variables in this case. But for $\sqcap[V]\notin[V]$, λ_x cannot be a Σ-substitution for any $x\in V$. We need a new variable, say y, and $\lambda_y=\{x_1\leftarrow y,\ldots,x_k\leftarrow y\}$ unifies V.

But which sort do we stipulate for y? We have to guarantee $[y]\in lbs[V]$, because otherwise we cannot expect the result to be well-sorted (which is a matter of correctness). As we shall see subsequently $[y]\in max(lbs[V])$ is a necessary condition to obtain a minimal set of Σ-unifiers. But max(lbs[V]) may have more than one element and as a matter of completeness we need a new variable y(s) with $[y(s)]=s$ for each $s\in max(lbs[V])$, to guarantee that $U=\{\lambda_{y(s)}\in SUB_\Sigma \mid s\in max(lbs[V])\}$ is a complete and minimal subset of $\mathcal{U}_\Sigma(V)$.

To define the weakening substitutions for a given $D\subset T_\Sigma$, we have to identify the subsets V of $\mathcal{V}(D)$ (as the set V above), which necessitate a variable replacement. The following technical lemma proves that these sets are given as those equivalence classes of $\underset{D}{\sim}$ of which the sortal image has no least element.

<u>Lemma 7.3</u>
For each $M\in T/\underset{D}{\sim}$, $M\subset\mathcal{V}(D)$ or $\sqcap[M]\in[M]$.

Proof For $M\subset\mathcal{V}(D)$ the lemma holds trivially, so let us assume $q\notin\mathcal{V}(D)$ for some $q\in M$.

Case (i) $q\in\mathcal{V}$: Then $q\notin DOM(\sigma)\cup\mathcal{V}(COD(\sigma))$ for each mgu σ of D, because $DOM(\sigma)\cup\mathcal{V}(COD(\sigma))\subset\mathcal{V}(D)$, cf. Lemma 3.2. Hence $\sigma q=\sigma r$ iff $q=r$ for each $r\in T$, i.e. $M=\{q\}$ and obviously $\sqcap[M]\in[M]$.∅

Case (ii) $q\notin\mathcal{V}$: For $\theta\in\mathcal{U}_\Sigma(D)$ θ unifies M, i.e. $\theta q=\theta r$ and in particular $[\theta q]=[\theta r]$ for each $r\in M$. With $\theta\in SUB_\Sigma$ we obtain $[\theta r]\leq[r]$, hence $[\theta q]\leq[r]$ and therefore $[\theta q]\in lbs[M]$. But $[\theta q]=[q]$ because $q\notin\mathcal{V}$, hence $[q]=\sqcap[M]\in[M]$.∅⊠

Note that by the above lemma, $\mathcal{M}=\{M\in T/_{\widetilde{D}} \mid \sqcap[M]\notin[M]\}$ is a finite set of finite subsets of $\mathcal{V}(D)$. This fact guarantees that the following definition, which renders our idea of how new variables have to be introduced precise, is meaningful.

<u>Definition 7.1</u>

Let $\mathcal{M}=\{M\in T/_{\widetilde{D}} \mid \sqcap[M]\notin[M]\}$. If $\mathcal{M}\neq\emptyset$, then by Lemma 7.3 $\mathcal{M}=\{V_1,\ldots,V_n\}$ for some $V_1,\ldots,V_n\subset\mathcal{V}(D)$ and we define for each i with $1\leq i\leq n$:

(1) $Z_i\subset\mathcal{V}\setminus\mathcal{V}(D)$ such that $\max(\mathrm{lbs}[V_i])\subset[Z_i]$ and $Z_i\cap Z_h=\emptyset$ for any h with $1\leq h\leq n$ and $h\neq i$,

(2) var_i is a mapping from $\max(\mathrm{lbs}[V_i])$ to Z_i such that $[var_i(s)]=s$ for all $s\in\max(\mathrm{lbs}[V_i])$, and

(3) $W_i=\{\mu_i^s\in SUB \mid \mu_i^s=\{x_i\leftarrow var_i(s)\}$ for some $s\in\max(\mathrm{lbs}[V_i])\}$, where $x_i\in V_i$ is arbitrary but fixed.

A most general weakening set or mgw-set $W(D)$ of D is a subset of SUB satisfying

(4) $W(D)=\{\mu\in SUB \mid \mu=\mu_1\circ\ldots\circ\mu_n$, where $\mu_i\in W_i\}$, where each element of $W(D)$ is called a weakening substitution.

Finally, the sortal degree $\Delta D\in\mathbb{N}$ of D is defined as

(5) $\Delta D=|\max(\mathrm{lbs}[V_1])\times\ldots\times\max(\mathrm{lbs}[V_n])|$.

For $\mathcal{M}=\emptyset$ we define $W(D)=\{\varepsilon\}$ and $\Delta D=1$. ⊠

Mgw-sets do exist for each Σ-unifiable Σ-termset D. They consist of Σ-substitutions and their cardinality is given by the sortal degree ΔD of D. For instance n=1, $V_1=D$, $Z_1=\{z_h \mid h\in S_k\}$ and $W_1=W(D)=\{\{x\leftarrow z_h\} \mid h\in S_k\}$ is an mgw-set of D as given in Example 7.2, where ΔD=k.

Lemma 7.4

An mgw-set of D exists and for each mgw-set W of D, $|W|=\Delta D$ and $W\subset SUB_\Sigma$.

Proof Let all notions be given as in Definition 7.1. If $\mathcal{M}=\emptyset$, then $W=\{\varepsilon\}\subset SUB_\Sigma$ is an mgw-set of D, W is the only mgw-set of D and $|W|=1=\Delta D$.

For $\mathcal{M}\neq\emptyset$ we obtain the existence of W_i by Lemma 7.3 and $|W_i|=|max(lbs[V_i])|$, hence $|W|=\Delta D$ and so W obviously exists.

Finally, for each member μ of an mgw-set of D, $\mu\mathcal{V}\subset\mathcal{V}\subset T_\Sigma$ and if $x\in DOM(\mu)$, then $x\in V_i$ and $[\mu x]\in lbs[V_i]$ for some i, i.e. $[\mu x]\leq[x]$. Hence $\mu\in SUB_\Sigma$ and $W\subset SUB_\Sigma$. ⊠

As we have seen at the beginning of this chapter, $\sqcap[V]\in[V]$ is a sufficient condition for $V\subset\mathcal{V}(D)$ to have a Σ-mgu. The introduction of auxiliary variables by a weakening substitution μ guarantees that this condition is satisfied for all equivalence classes of $\underset{\mu D}{\sim}$:

Lemma 7.5

Let μ be a member of an mgw-set of D. Then $\sqcap[N]\in[N]$ for each $N\in T/\underset{\mu D}{\sim}$.

Proof Since μ replaces different variables in D by new different variables, μD is unifiable and for each $N\in T/\underset{\mu D}{\sim}$ exactly one $M\in T/\underset{D}{\sim}$ exists such that $\mu M=N$. Let all notions be given as in Definition 7.1:

Case (i) $\sqcap[M]\in[M]$: Suppose that $[\mu M]\neq[M]$. Since $[\mu q]=[q]$ for each $q\in T\setminus\mathcal{V}$ some $x\in M\cap DOM(\mu)$ exists. Hence $x\in V_i$ for some $V_i\in T/\underset{D}{\sim}$ satisfying $\sqcap[V_i]\notin[V_i]$. But x is a member of exactly one equiva-

lence class of $\underset{D}{\sim}$, and therefore $M=V_i$, contradicting $\sqcap[M]\in[M]$. ∇
Hence $[\mu M]=[M]$ and thus $\sqcap[\mu M]\in[\mu M]$. $\boxtimes$

Case (ii) $\sqcap[M]\notin[M]$: Then with Definition 7.1 for some i and some $s_i\in\max(\mathrm{lbs}[V_i])$, $M=V_i$ and $[\mu M]\subset[V_i]\cup\{s_i\}$. But then $s_i\in\mathrm{lbs}([V_i]\cup\{s_i\})\subset\mathrm{lbs}[\mu M]$. Now let $s\in\mathrm{lbs}[\mu M]$. Then $s\leq s_i$ because $s_i\in[\mu M]$, hence $s_i=\sqcap[\mu M]\in[\mu M]$. $\boxtimes$

So in either case, $\sqcap[\mu M]\in[\mu M]$ and setting $N=\mu M$, the lemma has been proved. $\boxtimes$

<u>Lemma 7.6</u>

Let $\theta\in U_\Sigma(D)$ and let μ be a member of an mgw-set of D. Then for some $\tau\in SUB$

(1) τ is an mgu of μD, and
(2) $\tau\leq\theta[\![\mathcal{V}(D)\cap\mathcal{V}(\mu D)]\!]$.

Proof Let ρ be an mgu of D and let $\rho_1,\ldots,\rho_n\in SUB$ be a separation of ρ. We define $\tau_1,\ldots,\tau_n,\tau\in SUB$ by

(3) $DOM(\tau_i)=\mu DOM(\rho_i)$
(4) $COD(\tau_i)=\mu COD(\rho_i)$
(5) $DOM(\tau_n)=DOM(\rho_n)$
(6) $\tau_n=\mu\circ\rho_n[\![DOM(\tau_n)]\!]$, and
(7) $\tau=\tau_1\circ\ldots\circ\tau_n$,

where $1\leq i<n$. Since μ replaces different variables in D by new different variables, (1) obviously holds and

(8) $\bar{\mu}\circ\mu=\varepsilon[\![\mathcal{V}(D)]\!]$

for some $\bar{\mu}\in SUB$. Now let $x\in\mathcal{V}(D)\cap\mathcal{V}(\mu D)$. If $x\notin DOM(\tau)$, then $\theta\bar{\mu}\tau x=\theta\bar{\mu}x=\theta x$ and for $x\in DOM(\tau)$ we obtain $\theta\bar{\mu}\tau x=\theta\bar{\mu}\mu\rho_j x=\theta\rho_j x=\theta\rho x=\theta x$, because ρ is an mgu of D, θ unifies D and by (8). Hence in either case $\theta\circ\bar{\mu}\circ\tau=\theta[\![\mathcal{V}(D)\cap\mathcal{V}(\mu D)]\!]$ and (2) has been proved. $\boxtimes$

By Lemma 7.5 there is no more need of 'coercions' after an application of a weakening substitution, i.e all weakening substitutions $\mu \in W(D)$ share the interesting property that each μD has a Σ-mgu. This result will provide the key for the subsequent main theorems about many-sorted unification.

Theorem 7.7

Let μ be a member of an mgw-set of D. Then μD has a Σ-mgu.

Proof For each $\theta \in U_\Sigma(D)$ by Lemma 7.6 some $\tau \in SUB$ exists such that

(1) τ is an mgu of μD, and

(2) $\lambda \circ \tau = \theta [\![\mathcal{V}(D) \cap \mathcal{V}(\mu D)]\!]$

for some $\lambda \in SUB$. Let $\tau_1, \ldots, \tau_n$ be a separation of τ, let $\{y_i\} = COD(\tau_i)$ and let $V_i = DOM(\tau_i) \cup \{y_i\}$ for each i with $1 \le i < n$. Then $V_i \in T/_{\widetilde{\mu D}}$ for each i, because for any $u \in V_i$ and each $q \in T$ with $\tau u = \tau q$ we obtain $y_i = \tau q$, hence $q \in V_i$.

By Lemma 7.5 each $[V_i]$ has a least element, hence for some $x_i \in V_i, [x_i] = \sqcap [V_i]$. Let $\sigma_1 \ldots, \sigma_{n-1}, \sigma \in SUB$ be defined by

(3) $\sigma_i = \{y_i \leftarrow x_i\} \circ \tau_i$, for each i with $1 \le i < n$, and

(4) $\sigma = \sigma_1 \circ \ldots \circ \sigma_{n-1}$.

Since we obtain σ_i from τ_i by a variable renaming, we know $\sigma \circ \tau_n \equiv \tau$, hence

(5) $\sigma \circ \tau_n$ is an mgu of μD.

We show that

(6) $[\sigma_i x] \le [x]$, for each $x \in \mathcal{V}$ and each i.

For $x \notin V_i$ we have $\tau_i x = x \neq y_i$, i.e. $\sigma_i x = x$ and in particular $[\sigma_i x] = [x]$. If $x \in V_i$, then $\tau_i x = y_i$ and by (3) $\sigma_i x = x_i$. Hence $[\sigma_i x] = [x_i] \leq [x]$ because $[x_i] = \sqcap[V_i]$.

From (3) we infer $COD(\sigma_i) = \{x_i\}$, i.e. $\sigma_i x \in \mathcal{V} \subset T_\Sigma$ for each $x \in \mathcal{V}$, hence with (6) $\sigma_i \in SUB_\Sigma$ for each i, cf. Lemma 7.1, and by (4) we obtain

(7) $\sigma \in SUB_\Sigma$.

Now suppose $\tau_n x \in \mathcal{V}$ for some $x \in \mathcal{V}$. With $COD(\tau_n) \cap \mathcal{V} = \emptyset$ we obtain $\tau_n x = x$, i.e. $[\sigma\tau_n x] = [\sigma x] \leq [x]$ by (7). If $\tau_n x \notin \mathcal{V}$ for some $x \in \mathcal{V}$, then $x \in \mathcal{V}(D) \cap \mathcal{V}(\mu D)$ and we obtain $[\sigma\tau_n x] = [\tau_n x] = [\tau x] = [\lambda\tau x] = [\theta x] \leq [x]$ by (2) and $\theta \in SUB_\Sigma$. Hence in either case

(8) $[\sigma\tau_n x] \leq [x]$ for each $x \in \mathcal{V}$.

Next we prove that

(9) $\sigma\tau_n x \in T_\Sigma$ for each $x \in \mathcal{V}$.

For $x \notin DOM(\tau_n)$ we obtain $\sigma\tau_n x = \sigma x \in \mathcal{V}$ and with Lemma 4.3(1) (9) has been proved. Now let $x \in DOM(\tau_n)$ and $t = \tau_n x$. We assume by way of contradiction that $\sigma t \notin T_\Sigma$. Then by Lemma 4.3 t has a subterm $r = f(r_1 \ldots r_k)$ such that

(10) $[\sigma r_j] \nleq [f]_j$ for some j with $1 \leq j \leq k$.

From $x \in DOM(\tau_n)$ we obtain $x \in \mathcal{V}(D) \cap \mathcal{V}(\mu D)$, hence by (2) $\lambda t = \theta x$ and with $\theta \in SUB_\Sigma$, $\lambda t \in T_\Sigma$. Therefore $\lambda r \in T_\Sigma$ and in particular $[\lambda r_j] \leq [f]_j$.

Now if $r_j \notin \mathcal{V}$, then $[\sigma r_j] = [r_j] = [\lambda r_j] \leq [f]_j$, hence by (10)

(11) $r_j \in \mathcal{V}$.

Since r is a subterm of some member t of COD(τ) and τ is an mgu of μD, a subterm $q=f(q_1 \ldots q_k)$ of some term in μD exists such that

(12) $q \in T_\Sigma$, and
(13) $\tau q=r$.

Hence $\tau q_j=r_j$, therefore $\sigma r_j=\sigma\tau q_j$ and in particular

(14) $[\sigma r_j]=[\sigma\tau q_j]$.

By (11) we obtain $\tau q_j \in \mathcal{V}$, hence $q_j \in \mathcal{V}$ and $q_j \notin DOM(\tau_n)$ because $COD(\tau_n) \cap \mathcal{V}=\emptyset$.

If $q_j \notin V_i$ for each i then $\sigma\tau q_j=\sigma q_j=q_j$ and if $q_j \in V_i$ for some i, then $\sigma\tau q_j=\sigma\tau_i q_j=\sigma y_i=x_i$. With $[x_i]=\sqcap[V_i]$ we have $[\sigma\tau q_j]= [x_i] \le [q_j]$, hence in either case

(15) $[\sigma\tau q_j] \le [q_j]$.

From (12) we infer $[q_j] \le [f]_j$, hence by (14) and (15) $[\sigma r_j] \le [f]_j$ contradicting (10).∇

From (8) and (9) we obtain with Lemma 7.1 $\sigma \circ \tau_n \in SUB_\Sigma$, hence by (5) $\sigma \circ \tau_n$ is a Σ-mgu of D. ⊠

Now we can prove the existence of a complete and minimal set of Σ-unifiers for a Σ-unifiable Σ-termset: we define so called most general unifier sets and prove that each such set consists of Σ-unifiers, is complete and is also minimal.

<u>Definition 7.2</u>
Let W be some mgw-set of D. Then U(W) is a most general unifier set or mgu-set of D (relative to W) iff U(W) is a minimal subset of SUB satisfying: $\forall \mu \in W\ \exists \sigma \in SUB$ such that

(1) $\sigma \circ \mu \in U(W)$ and

(2) σ is a Σ-mgu of μD. ⊠

Mgu-sets do exist for each Σ-unifiable Σ-termset D. They consist of Σ-unifiers and their cardinality is given by the sortal degree ΔD of D. For instance $\{\{y \leftarrow z_h\} \circ \{x \leftarrow z_h\} \mid h \in S_k\} = \{\theta_h \mid h \in S_k\}$ is an mgu-set of D as given in Example 7.2.

<u>Lemma 7.8</u>

An mgu-set of D exists and for each mgu-set U of D, $|U| = \Delta D$ and $U \subset U_\Sigma(D)$.

Proof By Lemma 7.4 an mgw-set W of D exists and for each such set $|W| = \Delta D$ and $W \subset SUB_\Sigma$. By Theorem 7.7 for each $\mu \in W$ there is some $\sigma \in SUB_\Sigma$ such that σ is an mgu of μD. Hence for some $U \subset SUB$, U is an mgu-set of D.

Now let U be an mgu-set of D. Then by the minimality of U, $|U| = |W| = \Delta D$ and for each $\rho \in U$, $\rho = \sigma \circ \mu$ with $\sigma, \mu \in SUB_\Sigma$ and σ is an mgu of μD. Hence $\rho \in SUB_\Sigma$ and ρ unifies D, i.e. $\rho \in U_\Sigma(D)$. ⊠

To prove the completeness of an mgu-set we need the following technical lemma which states that each Σ-unifier is Σ-subsumed by some weakening substitution.

<u>Lemma 7.9</u>

Let $\theta \in U_\Sigma(D)$ and let W be an mgw-set of D. Then $\mu \leq_\Sigma \theta [\![V(D)]\!]$ for some $\mu \in W$.

Proof Let all notions be given as in Definition 7.1. For $\mathcal{M} = \emptyset$, we obtain $W = \{\varepsilon\}$ and the lemma holds trivially.

So let us assume that $\mathcal{M}\neq\emptyset$. With $\theta\in\mathcal{U}_\Sigma(D)$ we obtain for any i $\theta\in\mathcal{U}(V_i)$, hence $[\theta x]\in lbs[V_i]$ for each $x\in V_i$ because $\theta\in SUB_\Sigma$. Since $\mathcal{S}$ is finite, each non-empty subset S^* of $\mathcal{S}$ has a maximal element, i.e. $(S^*,>)$ satisfies the minimum condition (cf. [Cohn 1981]) and we can prove by Noetherian induction

(1) $\forall s\in S^*\ \exists s_m\in max(S^*).\ s\leq s_m$

and in particular

(2) $\forall s\in lbs[V_i]\ \exists s_m\in max(lbs[V_i]).\ s\leq s_m$,

hence $[\theta x]\leq s_i$ for some $s_i\in max(lbs[V_i])$. With $s_i\in max(lbs[V_i])$ we obtain $s_i\leq[\mu_i x]$ for each $x\in V_i$ and some $\mu_i\in W_i$ with $COD(\mu_i)=\{var_i(s_i)\}$. Hence

(3) $[\theta x]\leq[\mu_i x]$ for some $\mu_i\in W_i$ and each $x\in V_i$

and some $\mu=\mu_1\circ\ldots\circ\mu_n\in W$ exists such that $\mu_i\in W_i$ and each μ_i satisfies (3). Let $\mu_i=\{x_i\leftarrow z_i\}$, i.e. $\mu=\{x_1\leftarrow z_1,\ldots,x_n\leftarrow z_n\}$, where $\{z_1,\ldots,z_n\}\cap\mathcal{V}(D)=\emptyset$ and suppose $\theta_{|\mathcal{V}(D)}=\{y_1\leftarrow\theta y_1,\ldots,y_m\leftarrow\theta y_m\}$. We define $\lambda\in SUB$ by $\lambda=\{z_1\leftarrow\theta x_1,\ldots,z_n\leftarrow\theta x_n,y_1\leftarrow\theta y_1\ldots,y_m\leftarrow\theta y_m\}$. From (3) we obtain $[\theta x_i]\leq[z_i]$ and with $\theta\in SUB_\Sigma$ obviously $\lambda\in SUB_\Sigma$.

Now let $x\in\mathcal{V}(D)$. If $x\in DOM(\mu)$, then $x=x_i$ and $\lambda\mu x_i=\lambda z_i=\theta x_i$. For $x\notin DOM(\mu)$ we obtain $\lambda\mu x=\lambda x=\theta x$ because $x\notin COD(\mu)$. Hence in either case $\lambda\circ\mu=\theta[\![\mathcal{V}(D)]\!]$ and with $\lambda\in SUB_\Sigma$, $\mu\leq_\Sigma\theta[\![\mathcal{V}(D)]\!]$. ⊠

<u>Theorem 7.10</u> (Completeness Theorem)
Let U be an mgu-set of D. Then for each $\theta\in\mathcal{U}_\Sigma(D)$ some $\rho\in U$ exists such that $\rho\leq_\Sigma\theta[\![\mathcal{V}(D)]\!]$.

Proof Let U=U(W) be an mgu-set relative to some mgw-set W of D. By Lemma 7.9 $\lambda\circ\mu=\theta[\![\mathcal{V}(D)]\!]$ for some $\mu\in W$ and some $\lambda\in SUB_\Sigma$. μD has a Σ-mgu by Theorem 7.7, hence by Definition 7.2 $\sigma\circ\mu\in U$ for some Σ-mgu σ of μD.

With $\theta\in U_\Sigma(D)$ we have $\lambda\circ\mu\in U_\Sigma(D)$, i.e. $\lambda\in U_\Sigma(\mu D)$, hence $\lambda=\lambda\circ\sigma$ by Lemma 3.2 (3). But then $\lambda\circ\sigma\circ\mu=\lambda\circ\mu=\theta[\![\mathcal{V}(D)]\!]$, i.e. $\sigma\circ\mu\leq_\Sigma\theta[\![\mathcal{V}(D)]\!]$. ⊠

To show the minimality of an mgu-set again we need a technical lemma before proving the main theorem.

<u>Lemma 7.11</u>

Let μ be a member of an mgw-set of D and let σ be a Σ-mgu of μD. Then $[\sigma\mu x]=[\mu x]$ for each $x\in DOM(\mu)$.

Proof Let all notions be given as in Definition 7.1. From $x\in DOM(\mu)$ we infer $x\in V_i$ for some i. Let us assume by way of contradiction that $[\sigma\mu x]\neq[\mu x]$. Hence

(1) $[\sigma\mu x]<[\mu x]$

because $\sigma\in SUB_\Sigma$. Since σ is an mgu of μD and μ replaces different variables in D by different new variables, σ is an mgu of μV_i, ie. $\sigma\mu v=\mu v_i$ for some $v_i\in V_i$ and each $v\in V_i$. But then $\sigma\mu x=\mu v_i$ for some $v_i\in V_i$.

Using (1) we obtain $[\mu v_i]<[\mu x]$, hence $v_i\neq x$ and with Definition 7.1 (3) $v_i\notin DOM(\mu)$. But then $\mu v_i=v_i$ and thus

(2) $[v_i]<[\mu x]$.

But by Definition 7.1 (2,3) $[\mu x]\leq[v]$ for each $v\in V_i$, i.e. $[\mu x]\leq[v_i]$, and by (2) $[\mu x]<[\mu x]$.∇ Hence $[\sigma\mu x]=[\mu x]$ for each $x\in DOM(\mu)$ and the lemma has been proved. ⊠

Theorem 7.12 (Minimality Theorem)
Let U be an mgu-set of D and let $\rho_1, \rho_2 \in U$. Then $\rho_1 = \rho_2$, if $\rho_1 \leq_\Sigma \rho_2 [\![\mathcal{V}(D)]\!]$.

Proof Let us assume by way of contradiction that $\lambda \circ \rho_1 = \rho_2 [\![\mathcal{V}(D)]\!]$ for some $\lambda \in SUB_\Sigma$, but $\rho_1 \neq \rho_2$. By Definition 7.2 there is an mgw-set W of D, some $\mu_1, \mu_2 \in W$ and some $\sigma_1, \sigma_2 \in SUB_\Sigma$ such that for j=1,2

(1) $\rho_j = \sigma_j \circ \mu_j$, and
(2) σ_j is a Σ-mgu of $\mu_j D$.

For $\mu_1 = \mu_2$, $U \setminus \{\rho_j\}$ (for any $j \in \{1,2\}$) is also an mgu-set of D by Definition 7.2. But this is impossible by the minimality of U, hence $\mu_1 \neq \mu_2$.

Let all notions be given as in Definition 7.1. From $\mu_1 \neq \mu_2$ we obtain $\mu_1 x_i \neq \mu_2 x_i$ for some $x_i \in DOM(\mu_1) = DOM(\mu_2)$, hence

(3) $[\mu_1 x_i] \neq [\mu_2 x_i]$, by Definition 7.1
(4) $\{[\mu_1 x_i], [\mu_2 x_i]\} \subset max(lbs[V_i])$, by Definition 7.1, and
(5) $[\mu_2 x_i] \leq [\mu_1 x_i]$, because

$$\begin{aligned}
[\mu_2 x_i] &= [\sigma_2 \mu_2 x_i], && \text{by Lemma 7.11 using (2)} \\
&= [\rho_2 x_i] , && \text{by (1)} \\
&= [\lambda \rho_1 x_i] , && \text{by assumption} \\
&\leq [\rho_1 x_i] , && \text{since } \lambda \in SUB_\Sigma \\
&= [\sigma_1 \mu_1 x_i], && \text{by (1)} \\
&= [\mu_1 x_i] , && \text{by Lemma 7.11 using (2).}
\end{aligned}$$

From (4) and (5) we obtain $[\mu_1 x_i] = [\mu_2 x_i]$ which contradicts (3).
∇⊠

Now the Σ-Unification Theorem is an obvious consequence of the above lemmata:

Corollary 7.13 (Σ-Unification Theorem)
Let $D \subset T_\Sigma$ such that $\mathcal{U}_\Sigma(D) \neq \emptyset$. Then $mc\mathcal{U}_\Sigma(D) \neq \emptyset$ and $|mc\mathcal{U}_\Sigma(D)| = \Delta D$.

Proof By Lemma 7.8 an mgu-set $U \subset \mathcal{U}_\Sigma(D)$ of D exists. U is complete by Theorem 7.10 and minimal by Theorem 7.12, hence $mc\mathcal{U}_\Sigma(D) \neq \emptyset$. With $|U| = \Delta D$ by Lemma 7.8 and $|U_1| = |U_2|$ for each complete and minimal $U_1, U_2 \subset \mathcal{U}_\Sigma(D)$ (cf. Chapter 4) we finally obtain $|mc\mathcal{U}_\Sigma(D)| = \Delta D$. ⊠

Depending on the structure of $(\mathcal{S}, <_{\mathcal{S}})$ ΔD can be very large for a Σ-unifiable subset of T_Σ. Hence $U \in mc\mathcal{U}_\Sigma(D)$ can be very large, causing problems in computing and using U in an automated theorem prover. However, this problem can be avoided, if $(\mathcal{S}, <_{\mathcal{S}})$ is embedded into a meet-semilattice by the invention of additional sorts (cf. [McNeille 1937], Chapter 11 and also [Irani and Shin 1985]):

Corollary 7.14
If $(\mathcal{S}, <_{\mathcal{S}})$ is a meet-semilattice, then $|mc\mathcal{U}_\Sigma(D)| = 1$.

Proof If $(\mathcal{S}, <)$ is a meet-semilattice, $\sqcap S$ exists for each nonempty $S \subset \mathcal{S}$, i.e. $\max(lbs(S)) = \{\sqcap S\}$. Hence $\Delta D = 1$ and with Corollary 7.13 this corollary has been proved. ⊠

Finally we consider the special case where $(\mathcal{S}, <_{\mathcal{S}})$ is a forest structure, i.e. a finite collection of finite trees. This special case deserves our interest because we can prove that being a forest structure is a necessary and sufficient condition for a sort hierarchy such that the Robinson Unification Theorem holds for many-sorted unification. We need the following technical lemma about forest-structured sort hierarchies.

Lemma 7.15
Let $(\mathcal{S}, <_{\mathcal{S}})$ be a forest-structured sort hierarchy and let $S \subset \mathcal{S}$ such that $S \neq \emptyset$. Then $\sqcap S \in S$, if $lbs(S) \neq \emptyset$.

Proof If $lbs(S) \neq \emptyset$, then for some $s_o \in \mathcal{S}$, $s_o \leq s$ for each $s \in S$. Let $s_1, s_2 \in S$. Then $s_1 \geq s_o \leq s_2$, hence $s_1 \leq s_2$ or $s_1 \geq s_2$ because $(\mathcal{S}, <_{\mathcal{S}})$ is a forest. Therefore S is a chain in $\mathcal{S}$ and since S is finite and not empty $s' \leq s$ for some $s' \in S$ and each $s \in S$, i.e. $s' = \sqcap S \in S$. ⊠

Theorem 7.16

$(\mathcal{S}, <_{\mathcal{S}})$ is a forest-structured sort hierarchy iff each Σ-unifiable $D \subset T_\Sigma$ has a Σ-mgu.

Proof '→' Let $\theta \in U_\Sigma(D)$ and let $M \in T/\tilde{_D}$. Then θ unifies M, i.e. $\theta q = r$ for some $r \in T$ and each $q \in M$. With $\theta \in SUB_\Sigma$ we obtain $[r] = [\theta q] \leq [q]$ for each $q \in M$, hence $lbs([M]) \neq \emptyset$ and with Lemma 7.15 $\sqcap[M] \in [M]$. Hence $W = \{\varepsilon\}$ is the only mgw-set of D and by Theorem 7.7 $\varepsilon D = D$ has a Σ-mgu. ⊠

'←' Let $s(1), s(2), s \in \mathcal{S}$ such that

(*) $s(1) \geq s \leq s(2)$

and let $x_1 \in \mathcal{V}_{s(1)}$ $x_2 \in \mathcal{V}_{s(2)}$ and $y \in \mathcal{V}_s$. Then $\theta = \{x_1 \leftarrow y, x_2 \leftarrow y\}$ unifies $D = \{x_1, x_2\}$ and by (*), $\theta \in SUB_\Sigma$. By assumption D has a Σ-mgu, hence $\sigma_1 = \{x_1 \leftarrow x_2\}$ or $\sigma_2 = \{x_2 \leftarrow x_1\}$ is in SUB_Σ.

For $\sigma_1 \in SUB_\Sigma$ we obtain with Lemma 7.1 $s(1) = [x_1] \geq [\sigma_1 x_1] = [x_2] = s(2)$ and if $\sigma_2 \in SUB_\Sigma$, then $s(1) = [x_1] = [\sigma_2 x_2] \leq [x_2] = s(2)$. Hence $s(1) \geq s(2)$ or $s(1) \leq s(2)$ and therefore $(\mathcal{S}, <_{\mathcal{S}})$ is a forest structure. ⊠⊠

Note that Theorem 7.16 also holds for infinite sort hierarchies. However, Theorem 7.10 and consequently Corollary 7.13 do not hold in general, if $|\mathcal{S}| = \infty$ (cf. [Walther 1986]).

Obviously Corollary 7.13, Corollary 7.14 and Theorem 7.16 also hold, if D is a Σ-unifiable set of Σ-literals.

8 Completeness of the ΣRP-Calculus – The General Case

We will prove in this chapter the completeness theorem for the ΣRP-calculus in the general case. This theorem is shown in the same way as the completeness result for the one-sorted RP-calculus: we start with the Lifting Lemmata for Σ-Resolution and Σ-Paramodulation both of which depend heavily on the Σ-Unification Theorem. Using the Lifting Lemmata we can prove the Lifting Theorem for Σ-Deductions and with the Ground Completeness Theorem for ΣRP we finally obtain the desired completeness proof.

However, the completeness of the ΣRP-calculus is proved here only for sets of Σ-clauses, which contain all functionally-reflexive axioms [Wos and Robinson 1973]. For the RP-calculus it is known that these axioms are not a necessary prerequisite for the completeness results to hold [Brand 1975]. We conjecture that this result holds for the ΣRP-calculus also.

Although the approach to the completeness theorem is the same as for the RP-calculus, there are some essential differences caused by the differences between the definitions of resolution and paramodulation and their many-sorted versions.

We use members of a minimal and complete set of unifiers instead of a most general unifier where the completeness property of these sets guarantees that well-sorted ground deductions can always be lifted to Σ-deductions. The main difference, however, comes from the definition of Σ-paramodulation, because the use of coercion substitutions has no counterpart in the one-sorted case. But we do need these substitutions, because (as opposed to many-sorted resolution) using only well-sorted clauses and substitutions is not sufficient for the resulting paramodulant to be a well-sorted clause:

Let $C,D\in\mathcal{L}_\Sigma$ with $L\in C$, $E(qr)\in D$ and let $P=\sigma(C-L)\cup\sigma(D-E(qr))\cup\{\sigma K\}$, where $\sigma\in mcU_\Sigma(\{\alpha(L),q\})$ and $K\in LIT$ such that $\sigma L\sim_\alpha\sigma K$ and $\alpha(\sigma K)=\sigma r$ for some $\alpha\in SEL^+$. Then P is a well-sorted paramodulant iff $[\sigma r]\leq[L]_\alpha$, because $\sigma\in SUB_\Sigma$ guarantees that $\sigma(C-L)\cup\sigma(D-E(qr))$ and σL are well-sorted, and with $[\sigma r]\leq[L]_\alpha=[\sigma K]_\alpha$ the modulant literal σK is a well-sorted literal (cf. Lemma 4.2).

Hence an ad hoc solution to the problem of finding a reasonable definition for many-sorted paramodulation is to demand that $[\sigma r]\leq[L]_\alpha$ for each Σ-paramodulant (with r, L ,σ and α as given above). But unfortunately this restriction only works for the ground case (cf. Chapter 6), i.e. it is so strong that we lose completeness in the general case.

Example 8.1

Let $S=\{\{P(b_1)\},\{\neg P(b_2)\},\{E(xy)\}\}$ be a set of Σ-clauses such that $\mathcal{S}=\{A,B\}$, $B<_\mathcal{S}A$, $P\in\mathcal{P}_B$, $\{b_1,b_2\}\subset\mathcal{C}_B$ and $x,y\in\mathcal{V}_A$. Obviously each of the clauses $\{P(x)\}$, $\{\neg P(x)\}$, $\{P(y)\}$ and $\{\neg P(y)\}$ is not a Σ-clause and therefore cannot be obtained from S by Σ-paramodulation (with the restriction as given above). Hence $S\overset{E}{\vdash}_\Sigma C$ iff $C\in S$ and with $\Box\notin S$ we obtain $S\overset{E}{\nvdash}_\Sigma\Box$. But S is ΣE-unsatisfiable, because $S_{\Sigma gr}=\{\{P(b_1)\},\{\neg P(b_2)\},\{E(b_1b_2)\}, \ \ldots\ \}$ is E-unsatisfiable. ⊠

The problem here is that we replace in σL a term σq by a variable σr yielding a modulant literal σK with $\sigma L\sim_\alpha\sigma K$ and $\alpha(\sigma K)=\sigma r$ such that $[\sigma r]\nleq[L]_\alpha=[\sigma K]_\alpha$, i.e. $K\notin LIT_\Sigma$, but K has a well-sorted instance $\theta\sigma K$, i.e. $[\theta\sigma r]<[\sigma r]$ and $[\theta\sigma r]<[\sigma K]_\alpha$ for some $\theta\in SUB_\Sigma$.

Obviously this problem can arise only if σr is a variable, because we obtain with $\sigma r\notin\mathcal{V}$ that $[\theta\sigma r]=[\sigma r]\nleq[\sigma K]_\alpha$ for all $\theta\in SUB_\Sigma$. Hence a solution to the problem is to replace the variable σr in the paramodulant $P=\sigma(C-L)\cup\sigma(D-E(qr))\cup\{\sigma K\}$ by a new variable v satisfying $[v]<[\sigma r]$ and $[v]<[L]_\alpha$.

Obviously this replacement is a sound inference step. But as a matter of completeness, we need for each $s \in lbs(\{[\sigma r],[L]_\alpha\})$ a variable v_s such that $s \leq [v_s]$. As a matter of efficiency we have to keep the set of replacement variables as small as possible and choosing $v_{s'} \in \mathcal{V}_{s'}$ for all $s' \in max(lbs([\sigma r],[L]_\alpha))$ obviously satisfies both requirements.

But how do we find (and compute) these variable replacements? We map this problem into a many-sorted unification problem: technically, each of these variable replacements can be defined as an application of a well-sorted substitution (called coercion substitution) from the set of unifiers $U=mcU_\Sigma(\{\sigma r,z\})$, where $z \in \mathcal{V} \setminus \mathcal{V}(CUD)$ and $[z]=[L]_\alpha$. Let us consider the follwing cases (cf. Chapter 7):

Case 1 $U=\emptyset$: Then $lbs(\{[\sigma r],[L]_\alpha\})=\emptyset$, i.e. the modulant σK cannot have a well-sorted instance, and hence there is no Σ-paramodulant and no coercion substitution in this case.

Case 2 $U \neq \emptyset$ and $[\sigma r]<[L]_\alpha$: Then σK is a modulant literal, which is already well-sorted, and we do not need a variable replacement. But with $U=\{\{z \leftarrow \sigma r\}\}$ and $z \notin \mathcal{V}(P)$ we have $\{z \leftarrow \sigma r\}P=P$ and consequently there is no variable replacement.

Case 3 $U \neq \emptyset$ and $[\sigma r]>[L]_\alpha$: Then $\sigma r \in \mathcal{V}$, $U=\{\nu\}$ with $\nu=\{\sigma r \leftarrow z\}$ and νP is P with σr replaced by z thus guaranteeing that $\nu\sigma K$ is a well-sorted literal.

Case 4 $U \neq \emptyset$ and $[\sigma r]=[L]_\alpha$: Then $U=\{\nu\}$ with $\nu=\{\sigma r \leftarrow z\}$ (where $\sigma r \in \mathcal{V}$) or $\nu=\{z \leftarrow \sigma r\}$. As in case 2, σK is a modulant literal, which is already well-sorted, and we do not need a variable replacement. With $\nu=\{\sigma r \leftarrow z\}$ we replace σr by a new variable z of the same sort (which of course is an unnecessary operation) and with $\nu=\{z \leftarrow \sigma r\}$ the variable replacement has no effect as in case 2.

Case 5 $U \neq \emptyset$, $[\sigma r] \not\leq [L]_\alpha$ and $[\sigma r] \not\geq [L]_\alpha$: Then $\sigma r \in \mathcal{V}$, $U=\{\{\sigma r \leftarrow v_1, z \leftarrow v_1\}, \{\sigma r \leftarrow v_2, z \leftarrow v_2\}, \ldots, \{\sigma r \leftarrow v_k, z \leftarrow v_k\}\}$ where $V=\{v_1, v_2, \ldots, v_k\}$ is a minimal subset of $\mathcal{V}$ satisfying $[V]=\max(\mathrm{lbs}(\{[\sigma r], [L]_\alpha\}))$. For each v_i, $v_i P$ is P with σr replaced by v_i and with $v_i \in SUB_\Sigma$ we have $[v_i]=[v_i \sigma r] \leq [\sigma r]$ and $[v_i]=[v_i z] \leq [z]=[\sigma K]_\alpha$, hence $v\sigma K$ is a well-sorted instance of the modulant literal σK.

This discussion motivates our definition of many-sorted paramodulation as given in Chapter 4. For instance, for the set of Σ-clauses S as given in Example 8.1, we obtain P(z) as a Σ-paramodulant of $P(b_1)$ and E(xy) (with coercion substitution $v=\{y \leftarrow z\}$, where $z \in \mathcal{V}_B$, cf. case 3) and $\square$ as a Σ-resolvent of P(z) and $\neg P(b_2)$.

However, with this definition Σ-Paramodulation is so restrictive that we lose an important refinement: the RP-calculus is still complete if we never paramodulate into positive equality literals [Loveland 1978]. The consequence is that the transitive closure of the predicate E need not be computed, i.e. in the RP-calculus we do not need deductions of the form: $\{\{E(qr)\}, \{E(rs)\}\} \vdash_P \{E(qs)\}$. But unfortunately these deductions are necessary in the ΣRP-calculus:

<u>Example 8.2</u>

Let $S=\{\{P(b_1)\}, \{\neg P(b_2)\}, \{E(b_1 a)\}, \{E(a b_2)\}\}$ be a set of Σ-clauses such that $\mathcal{S}=\{A,B\}$, $A \not\leq_{\mathcal{S}} B$, $P \in \mathcal{P}_B$, $\{b_1, b_2\} \subset \mathcal{C}_B$ and $a \in \mathcal{C}_A$. Obviously neither $\{P(a)\}$ nor $\{\neg P(a)\}$ is a Σ-clause, i.e. they can not be obtained from S by Σ-paramodulation. But with the Σ-paramodulant $\{E(b_1 b_2)\}$ of $\{E(b_1 a)\}$ and $\{E(a b_2)\}$ we obtain the Σ-paramodulants $\{P(b_2)\}$ and $\{\neg P(b_1)\}$ each of which leads to a Σ-refutation of S. ⊠

In the general case, it may even become necessary to paramodulate a clause with itself:

Example 8.3

Let $S=\{\{P(b_1)\},\{\neg P(b_2)\},\{E(xa)\}\}$ be a set of Σ-clauses such that $\mathcal{S}=\{A,B\}$, $B<_{\mathcal{S}}A$, $P\in\mathcal{P}_B$, $x\in\mathcal{V}_A$ and $a\in\mathcal{C}_A$. Obviously neither $\{P(a)\}$ nor $\{\neg P(a)\}$ is a Σ-clause, i.e. they can not be obtained from S by Σ-paramodulation. But with the Σ-paramodulant $\{E(xy)\}$ of $\{E(xa)\}$ and the Σ-renamed variant $\{E(ya)\}$ of itself (where $y\in\mathcal{V}_A$ is a fresh variable) we obtain the Σ-paramodulants $\{P(u)\}$ and $\{\neg P(v)\}$ (where $u,v\in\mathcal{V}_B$ are fresh variables) each of which leads to a Σ-refutation of S. ⊠

The problem is the transitive closure of E, hence we conjecture that Σ-paramodulation is still complete if we never paramodulate into subterms of positive equality literals unless these subterms are in an argument position of an equality literal.

Now we present the formal proofs. To ease notation, we do not mention Σ-renaming substitutions in Σ-deductions explicitly. Instead we will assume that all clauses in a Σ- deduction share no variable symbols.

Lemma 8.1 (Lifting Lemma for Σ-Resolution)

Given $A,B\in\mathcal{L}_\Sigma$, $L_A\in A$, $L_B\in B$ and $\theta\in SUB_\Sigma$ such that A and B share no variable symbols, L_A and L_B are complementary and θ unifies $\{|L_A|,|L_B|\}$, there exist a Σ-factor A^* of A, a Σ-factor B^* of B, a pair of complementary literals $L_A^*\in A^*$ and $L_B^*\in B^*$, and some $\lambda,\sigma\in SUB_\Sigma$ such that

$$Res(\theta A,\theta L_A,\theta B,\theta L_B,\varepsilon) = \lambda Res_\Sigma(A^*,L_A^*,B^*,L_B^*,\sigma).$$

The statement of this Lemma can be summarized by a diagram. The expressions in the diagram are as defined in the Lemma:

$$
\begin{array}{ccc}
A,B & \xrightarrow{F+R} & Res_{\Sigma}(A^*,L_A^*,B^*,L_B^*,\sigma) \\
\Big\downarrow \Sigma\text{-instance} & & \Big\downarrow \Sigma\text{-instance} \\
 & & \lambda Res_{\Sigma}(A^*,L_A^*,B^*,L_B^*,\sigma) \\
 & & = \\
\theta A,\theta B & \xrightarrow{R} & Res(\theta A,\theta L_A,\theta B,\theta L_B,\varepsilon)
\end{array}
$$

Figure 8.1 Lifting Lemma for Σ-Resolution

Proof Let $A'=\{L\in A \mid \theta L=\theta L_A\}$. Then obviously $\theta\in U_{\Sigma}(A')$ and by the Σ-Unification Theorem (7.13) some $\gamma\in mcU_{\Sigma}(A')$ exists such that

(1) $\theta=\lambda_A\circ\gamma[\![\mathcal{V}(A')]\!]$

for some $\lambda_A\in SUB_{\Sigma}$. Obviously γA is a Σ-factor of A and by an analogous argument

(2) $\theta=\lambda_B\circ\delta[\![\mathcal{V}(B')]\!]$

for some $\lambda_B\in SUB_{\Sigma}$ and some $\delta\in SUB_{\Sigma}$ which factors B.

Since we are allowed to assume that γA and δB have no variables in common by (1) and (2) there exists some $\lambda^*\in SUB_{\Sigma}$ such that

(3) $\lambda^*=\lambda_A[\![\mathcal{V}(\gamma A)]\!]$, and

(4) $\lambda^*=\lambda_B[\![\mathcal{V}(\delta B)]\!]$.

Let $A^*=\gamma A$, $B^*=\delta B$, $V^*=\mathcal{V}(A^*\cup B^*)$ and L^* be an abbreviation for γL if $L\in A$ and for δL if $L\in B$. Then for each $L\in A$, $\lambda^*L^*=\lambda^*\gamma L=\lambda_A\gamma L=\theta L$ by (3) and (1), and by analogous argument $\lambda^*L^*=\theta L$ for each $L\in B$, i.e.

(5) $\lambda^*L^*=\theta L$, for each $L\in(A\cup B)$.

Hence $\lambda^*|L_A^*|=\theta|L_A|=\theta|L_B|=\lambda^*|L_B^*|$, i.e. λ^* is a Σ-unifier of $\{|L_A^*|,|L_B^*|\}$, and by the Σ-Unification Theorem (7.13) some $\sigma\in mc\mathcal{U}_\Sigma(\{|L_A^*|,|L_B^*|\})$ exists such that

(6) $\lambda^*=\lambda\circ\sigma[\![V^*]\!]$

for some $\lambda\in SUB_\Sigma$ and obviously the Σ-resolvent $Res_\Sigma(A^*,L_A^*,B^*,L_B^*,\sigma)$ exists.

Now suppose that $\lambda^*(A^*-L_A^*)\neq\lambda^*A^*-\lambda^*L_A^*$. Then for some $L^*\in A^*$ we have $\lambda^*L^*=\lambda^*L_A^*$ but $L^*\neq L_A^*$. From (5) we obtain $\theta L=\theta L_A$, hence $L\in A'$. But then $\gamma L=\gamma L_A$ because γ Σ-unifies A', i.e. $L^*=L_A^*$. ∇

Thus we have proved that

(7) $\lambda^*(A^*-L_A^*)=\lambda^*A^*-\lambda^*L_A^*$

and we obtain by similar argument

(8) $\lambda^*(B^*-L_B^*)=\lambda^*B^*-\lambda^*L_B^*$.

But then

$$
\begin{aligned}
&\lambda Res(A^*,L_A^*,B^*,L_B^*,\sigma) && ,\\
&\quad = \lambda\sigma(A^*-L_A^*)\cup\lambda\sigma(B^*-L_B^*) && ,\\
&\quad = \lambda^*(A^*-L_A^*)\cup\lambda^*(B^*-L_B^*) && ,\ \text{by (6)}\\
&\quad = (\lambda^*A^*-\lambda^*L_A^*)\cup(\lambda^*B^*-\lambda^*L_B^*) && ,\ \text{by (7) and (8)}\\
&\quad = (\theta A-\theta L_A)\cup(\theta B-\theta L_B) && ,\ \text{by (5)}\\
&\quad = Res(\theta A,\theta L_A,\theta B,\theta L_B,\varepsilon) && .\ \boxtimes
\end{aligned}
$$

<u>Lemma 8.2</u> (Lifting Lemma for Σ-Paramodulation)
Given $A,B\in\mathcal{L}_\Sigma$, $L_A\in A$, $L_B=E(qr)\in B$, $\alpha\in SEL^+$ and $\theta\in SUB_\Sigma$ such that A and B share no variable symbols, θ unifies $\{\alpha(L_A),q\}$ and $[\theta r]\le[L_A]_\alpha$, there exist a Σ-factor A* of A, a Σ-factor B* of B, a pair of literals $L_A^*\in A^*$ and $L_B^*=E(q^*r^*)\in B^*$, and some $\rho,\nu,\sigma\in SUB_\Sigma$ such that

$$Par(\theta A,\theta L_A,\theta B,\theta L_B,\alpha,\varepsilon) = \rho Par_\Sigma(A^*,L_A^*,B^*,L_B^*,\alpha,\nu\circ\sigma).$$

The statement of this Lemma can be summarized by a diagram. The expressions in the diagram are as defined in the Lemma:

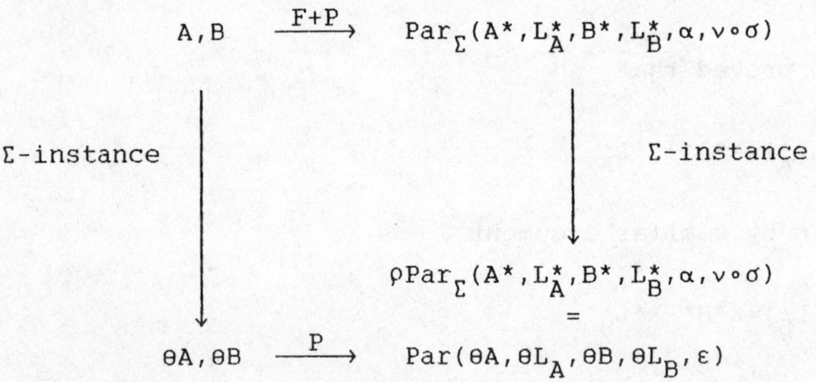

Figure 8.2 Lifting Lemma for Σ-Parmodulation

Proof By the same argument as in the proof of Lemma 8.1 there exist $\gamma,\delta,\lambda^*,\lambda,\sigma\in SUB_\Sigma$ such that $A^*=\gamma A$ is a Σ-factor of A, $B^*=\delta B$ is a Σ-factor of B,

(1) $\lambda^*L^*=\theta L$, for each $L\in(A\cup B)$,

where L* abbreviates γL or δL depending on L∈A or L∈B (and in particular $q^*=\delta q$, $r^*=\delta r$),

(2) $\sigma \in mcU_{\Sigma}(\{\alpha(L_A^*), q^*\})$,

(3) $\lambda^* = \lambda \circ \sigma[\![V^*]\!]$, for $V^* = \mathcal{V}(A^* \cup B^*)$,

(4) $\lambda^*(A^* - L_A^*) = \lambda^* A^* - \lambda^* L_A^*$, and

(5) $\lambda^*(B^* - L_B^*) = \lambda^* B^* - \lambda^* L_B^*$.

Let $K, K' \in LIT$ such that $L_A \underset{\alpha}{\sim} K$, $\alpha(K) = r$, $L_A^* \underset{\alpha}{\sim} K'$ and $\alpha(K') = r^*$. Then by Lemma 3.3 (4,5)

(6) $\theta L_A \underset{\alpha}{\sim} \theta K$ and $\alpha(\theta K) = \theta r$,

(7) $\sigma L_A^* \underset{\alpha}{\sim} \sigma K'$ and $\alpha(\sigma K') = \sigma r^*$,and

(8) $\lambda^* L_A^* \underset{\alpha}{\sim} \lambda^* K'$ and $\alpha(\lambda^* K') = \lambda^* r^*$.

Using (1) we obtain from (8) $\theta L_A \underset{\alpha}{\sim} \lambda^* K'$ and $\alpha(\lambda^* K') = \theta r$, and with (6) $\lambda^* K' \underset{\alpha}{\sim} \theta K$ and $\alpha(\lambda^* K') = \alpha(\theta K)$. Hence by Lemma 3.3 (1)

(9) $\lambda^* K' = \theta K$.

Now let $z \in \mathcal{V} \setminus \mathcal{V}(\sigma V^*)$ such that $[z] = [L_A]_{\alpha}$. We show, that for some $\nu \in mcU_{\Sigma}(\{z, \sigma r^*\})$ and some $\rho \in SUB_{\Sigma}$

(10) $\rho \circ \nu \circ \sigma = \lambda \circ \sigma[\![V^*]\!]$.

Case (i) $[\sigma r^*] \leq [z]$ and $\sigma r^* \notin \mathcal{V}$: Then $\nu = \{z \leftarrow \sigma r^*\} \in mcU_{\Sigma}(\{z, \sigma r^*\})$ and we define $\rho = \lambda$. For $y \in \mathcal{V}(\sigma V^*)$ we obtain $y \notin DOM(\nu)$, hence $\rho \nu y = \rho y = \lambda y$ for all $y \in \mathcal{V}(\sigma V^*)$ and therefore $\rho \nu \sigma x = \rho \sigma x = \lambda \sigma x$ for all $x \in \mathcal{V}(V^*)$. ☒

Case (ii) $[\sigma r^*] \nleq [z]$ or $\sigma r^* \in \mathcal{V}$: Then $\sigma r^* \in \mathcal{V}$ in either case, because we obtain with $\sigma r^* \notin \mathcal{V}$ that $[\sigma r^*] = [\lambda \sigma r^*]$ (by Lemma 4.1 (2)), $\ldots = [\lambda^* r^*] = [\theta r]$ (by (3) and (1)), and since $[\theta r] \leq [L_A]_{\alpha} = [z]$ by the hypothesis, we finally infer $[\sigma r^*] \leq [z]$. ∇

Let $\tau \in SUB$ be given as

(11) $\tau=\{z \leftarrow \theta r, \sigma r^* \leftarrow \theta r\}$.

Since $[\theta r] \leq [z]$ by the hypothesis and $[\theta r]=[\lambda^* r^*]=[\lambda \sigma r^*] \leq [\sigma r^*]$, we have $\tau \in SUB_\Sigma$. Obviously $\tau \in \mathcal{U}_\Sigma(\{z,\sigma r^*\})$ and by the Σ-Unification Theorem (7.13) there exist $\mu \in SUB_\Sigma$ and $\nu \in mc\mathcal{U}_\Sigma(\{z,\sigma r^*\})$ such that

(12) $\mu \circ \nu = \tau [\![\{\sigma r^*, z\}]\!]$

With $z, \sigma r^* \in \mathcal{V}$ and $\nu \in mc\mathcal{U}_\Sigma(\{z,\sigma r^*\})$, obviously $\nu \sigma r^* \in \mathcal{V}$ and we can define $\rho \in SUB$ as

(13) $\rho=\{\nu \sigma r^* \leftarrow \lambda^* r^*\} \circ \lambda$

Since $\lambda^* r^* = \theta r = \tau \sigma r^* = \mu \nu \sigma r^*$ by (1), (11) and (12), we have $[\lambda^* r^*] \leq [\nu \sigma r^*]$ and with $\lambda \in SUB_\Sigma$ we obtain $\rho \in SUB_\Sigma$. Now let $y \in \mathcal{V}(\sigma V^*)$ and suppose $y \in DOM(\nu)$, then $y=\sigma r^*$ and we have with (13) $\rho \nu y = \rho \nu \sigma r^* = \lambda^* r^* = \lambda \sigma r^* = \lambda y$. For $y \notin DOM(\nu)$ we know $y \neq \sigma r^*$ and therefore $y = \nu y \neq \nu \sigma r^*$. Hence with (13) $\rho \nu y = \rho y = \lambda y$. Hence in any case $\rho \circ \nu = \lambda [\![\mathcal{V}(\sigma V^*)]\!]$ and therefore $\rho \circ \nu \circ \sigma = \lambda \circ \sigma [\![V^*]\!]$. ☐

But then

$$\begin{aligned}
\rho Par_\Sigma(A^*, L_A^*, B^*, L_B^*, \alpha, \nu \circ \sigma) \\
&= \rho \nu \sigma (A^*-L_A^*) \cup \rho \nu \sigma (B^*-L_B^*) \cup \{\rho \nu \sigma K'\} && \text{, by (7),} \\
&= \lambda \sigma (A^*-L_A^*) \cup \lambda \sigma (B^*-L_B^*) \cup \{\lambda \sigma K'\} && \text{, by (10),} \\
&= \lambda^* (A^*-L_A^*) \cup \lambda^* (B^*-L_B^*) \cup \{\lambda^* K'\} && \text{, by (3),} \\
&= (\lambda^* A^* - \lambda^* L_A^*) \cup (\lambda^* B^* - \lambda^* L_B^*) \cup \{\lambda^* K'\} && \text{, by (4) and (5),} \\
&= (\theta A - \theta L_A) \cup (\theta B - \theta L_B) \cup \{\theta K\} && \text{, by (1) and (9),} \\
&= Par(\theta A, \theta L_A, \theta B, \theta L_B, \alpha, \varepsilon) && \text{, by (6).} \; \boxtimes
\end{aligned}$$

Theorem 8.3 (Lifting Theorem for Σ-Deductions)
Let $S \subset \mathcal{L}_\Sigma$ and $C \in \mathcal{L}_{\Sigma gr}$ such that $S^E_{\Sigma gr} \vdash_\Sigma C$. Then there exist $D \in \mathcal{L}_\Sigma$ and $\theta \in SUB_\Sigma$ such that $S^E \vdash_\Sigma D$ and $\theta D = C$.

Proof The proof is by induction upon the length n of the Σ-deduction of C from $S^E_{\Sigma gr}$:

Base case n=1: Then $C \in S^E_{\Sigma gr}$ and therefore $C=\theta D$ for some $D \in S^E$ and some $\theta \in SUB_\Sigma$. With $D \in S^E$ obviously $S^E \vdash_\Sigma D$. ☐

Induction step n>1: Our induction hypothesis is to assume that the theorem holds for all Σ-deductions from $S^E_{\Sigma gr}$ with length less than n.

Case (i) $C \in S^E_{\Sigma gr}$: For this case the theorem holds by the same argument as in the base case. ☐

Case (ii) $C=Res_\Sigma(C_i,L_i,C_j,L_j,\varepsilon)$ or $C=Par_\Sigma(C_i,L_i,C_j,L_j,\alpha,\nu\circ\varepsilon)$: Then by the induction hypothesis there exist $D_h, D_k \in \mathcal{L}_\Sigma$, θ_h, $\theta_k \in SUB_\Sigma$ such that $S^E \vdash_\Sigma D_h$, $S^E \vdash_\Sigma D_k$, $C_i=\theta_h D_h$ and $C_j=\theta_k D_k$. Since we are allowed to assume that D_h and D_k share no variable symbols, there exists $\theta^* \in SUB_\Sigma$ such that $\theta^*=\theta_h[\![\mathcal{V}(D_h)]\!]$ and $\theta^*=\theta_k[\![\mathcal{V}(D_k)]\!]$.

Case (ii.1) $C=Res_\Sigma(C_i,L_i,C_j,L_j,\varepsilon)$: Since there is no difference between resolution and Σ-resolution in the Σ-ground case (cf. Chapter 6) we know that $C=Res(C_i,L_i,C_j,L_j,\varepsilon)$ and by the Lifting Lemma for Σ-Resolution (8.1) there exist Σ-factors D^*_h and D^*_k of D_h and D_k respectively and a Σ-resolvent D of D^*_h and D^*_k such that $\theta D=C$ for some $\theta \in SUB_\Sigma$. ☐

Case (ii.2) $C=Par_\Sigma(C_i,L_i,C_j,L_j,\alpha,\nu\circ\varepsilon)$: Then $C=Par(C_i,L_i,C_j,L_j,\alpha,\varepsilon)$ (cf. Chapter 6). For $L_j=E(qr)$ we have $[r] \le [L_i]_\alpha$ because $S^E_{\Sigma gr} \vdash_\Sigma C$. Since S^E contains a functionally-reflexive axiom for each function symbol in C_i we are allowed to assume that $\alpha(L'_h)\downarrow$ for some $L'_h \in D_h$ with $\theta^* L'_h = L_i$.

Hence the hypotheses of the Lifting Lemma for Σ-Paramodulation (8.2) are satisfied and there exist Σ-factors D_h^* and D_k^* of D_h and D_k respectively and a Σ-paramodulant D of D_h^* and D_k^* such that $\theta D=C$ for some $\theta\in SUB_\Sigma$. ⊠ ⊠ ⊠ ⊠

Now we are prepared to prove the completeness of the ΣRP-calculus:

<u>Theorem 8.4</u> (Completeness Theorem for ΣRP)
If $S\subset\mathcal{L}_\Sigma$ is ΣE-unsatisfiable, then $S \overset{E}{\vdash}_\Sigma \square$.

Proof If S is ΣE-unsatisfiable, then $S_{\Sigma gr}$ is E-unsatisfiable and $S_{\Sigma gr} \overset{E}{\vdash}_\Sigma \square$ by the Ground Completeness Theorem for ΣRP (6.6). By the Lifting Theorem for Σ-Deductions (8.3), there exist $D\in\mathcal{L}_\Sigma$ and $\theta\in SUB_\Sigma$ such that $S \overset{E}{\vdash}_\Sigma D$ and $\theta D=\square$. Hence $D=\square$ and $S \overset{E}{\vdash}_\Sigma \square$ is proved. ⊠

9 Soundness of the ΣRP-Calculus

We will prove here the soundness theorem for the ΣRP-calculus. This is shown in the same way as the soundness theorem for the one-sorted RP-calculus [Loveland 1978]: first we prove with the soundness lemma that each ΣE-model of a Σ-clause set satisfies each clause deduced from this set by Σ-factorization, Σ-resolution or Σ-paramodulation. Then the soundness proof is obtained from the soundness lemma as an immediate consequence.

For the proof of the soundness lemma we use the following technical lemma, which exploits a trivial but essential circumstance regarding soundness:

Lemma 9.1

Let $A,B\in\mathcal{L}_\Sigma$ and $\theta\in SUB_\Sigma$ such that $A\subset B$ and $\theta A\in\mathcal{L}_{\Sigma gr}$. Then there exists some $\lambda\in SUB_\Sigma$ such that

(1) $\theta=\lambda[\![\mathcal{V}(A)]\!]$, and

(2) $\lambda B\in\mathcal{L}_{\Sigma gr}$.

Proof For $\mathcal{V}(B)\subset\mathcal{V}(A)$ we define $\lambda=\theta$ and the lemma holds trivially. Now assume $\mathcal{V}(B)\not\subset\mathcal{V}(A)$, let $\{x_1,\ldots,x_n\}=\mathcal{V}(B)\setminus\mathcal{V}(A)$ and let $\lambda,\delta\in SUB_\Sigma$ such that $\lambda=\theta\circ\delta$ where $\delta=\{x_1\leftarrow t_1,\ldots,x_n\leftarrow t_n\}$ for some $t_i\in T_{\Sigma gr}$ with $[t_i]\le[x_i]$ and $1\le i\le n$. It is obvious that λ satisfies conditions (1) and (2). ⊠

However, the crucial point in the proof is the existence of the above terms t_i. This is guaranteed only because we demand that for each sort symbol s which is minimal in $(\mathcal{S},<_{\mathcal{S}})$ there must exist a constant symbol $c\in\mathcal{C}_s$ (cf. Chapter 4). If this requirement is not fulfilled, i.e. there do exist "empty" sorts, then this lemma and in turn the results of this chapter do not hold. The ΣRP-calculus is then not sound, because for a non-empty Σ-clause set S, $S_{\Sigma gr}$ may be empty.

But as a matter of fact, the requirement that $\mathcal{C}_s \neq \emptyset$ for each minimal $s \in \mathcal{S}$ is only a generalization of a condition which guarantees the soundness of the RP-calculus: one demands that $\mathcal{C} \neq \emptyset$ thus preventing $S_{gr} = \emptyset$ for any non-empty clause set S, cf. [Chang and Lee 1973, Loveland 1978]. Consequently, the RP-calculus is sound. Note that the RP-calculus can be considered as a ΣRP-calculus for which $\mathcal{S}=\{s\}$, i.e. a singleton. Hence $\mathcal{C}=\mathcal{C}_s$ in this case and since s is minimal in $(\{s\}, <_{\{s\}})$ we have $\mathcal{C} \neq \emptyset$. Therefore the soundness requirement for the ΣRP-calculus entails the soundness requirement for the unsorted case.

Since the approach to the soundness theorem is the same as for the RP-calculus, the remainder of this chapter is included only for completeness of the presentation.

To ease notation, we do not mention Σ-renaming substitutions in Σ-deductions explicitly. Instead we will assume subsequently that all clauses in a Σ-deduction share no variable symbols.

<u>Lemma 9.2</u> (Soundness Lemma for ΣRP)
Let $S \subset \mathcal{L}_\Sigma$, $C \in \mathcal{L}_\Sigma$ such that $S \vdash_\Sigma C$ and let M be a ΣE-model of S. Then M Σ-satisfies C.

Proof The proof is by induction on the length n of the Σ-deduction of C from S.

Base Case n=1: Then $C \in S$ and M Σ-satisfies C by the hypothesis. $\square$

Induction Step n>1: Let $C_n \in \mathcal{L}_\Sigma$ such that the length of the Σ-deduction $S \vdash_\Sigma C_n$ is n. If $C_n \in S$, then M Σ-satisfies C_n by the same argument as in the base case. Now suppose that $C_n \notin S$.

Our induction hypothesis is to assume that M Σ-satisfies each Σ-clause C_i with $S \vdash_\Sigma C_i$, where $i<n$. Let $\theta \in SUB_\Sigma$ such that $\theta C_n \in \mathcal{L}_{\Sigma gr}$. We prove that $M \cap \theta C_n \neq \emptyset$:

Case (i) Factoring: Let $C_n = \sigma C_i$ such that $i<n$, $S \vdash_\Sigma C_i$ and C_n is a Σ-factor of C_i.

Then $\theta\sigma C_i \in \mathcal{L}_{\Sigma gr}$ and $M \cap \theta C_n = M \cap \theta\sigma C_i \neq \emptyset$ by the induction hypothesis. ∅

Case (ii) Resolution: Let $C_n = Res_\Sigma(C_i, L_i, C_j, L_j, \sigma)$ such that $i,j<n$, $S \vdash_\Sigma C_i$, $S \vdash_\Sigma C_j$ and $\sigma \in SUB_\Sigma$.

Since $C_n \subset \sigma(C_i \cup C_j)$, by Lemma 9.1 there exists some $\lambda \in SUB_\Sigma$ such that $\theta C_n = \lambda C_n$ and $\lambda\sigma(C_i \cup C_j) \in \mathcal{L}_{\Sigma gr}$.

Therefore $\lambda\sigma C_i, \lambda\sigma C_j \in \mathcal{L}_{\Sigma gr}$ and we obtain from the induction hypothesis that $M \cap \lambda\sigma C_i \neq \emptyset$ and $M \cap \lambda\sigma C_j \neq \emptyset$.

Case (ii.1) $\lambda\sigma L_i \notin M$ or $\lambda\sigma L_j \notin M$: If $\lambda\sigma L_i \notin M$ then $M \cap (\lambda\sigma C_i - \lambda\sigma L_i) \neq \emptyset$ and since $(\lambda\sigma C_i - \lambda\sigma L_i) \subset \lambda\sigma(C_i - L_i) \subset \theta C_n$ we have $M \cap \theta C_n \neq \emptyset$. With $\lambda\sigma L_j \notin M$ we obtain the same result. ∅

Case (ii.2) $\lambda\sigma L_i \in M$ and $\lambda\sigma L_j \in M$: With $\sigma L_i = \sigma L_j^c$ we obtain $\lambda\sigma L_i = \lambda\sigma L_j^c$, hence $\{\lambda\sigma L_j^c, \lambda\sigma L_j\} \in M$ and therefore M is not an interpretation. ∇ ∅ ∅

Case (iii) Paramodulation: Let $C_n = Par_\Sigma(C_i, L, C_j, E(qr), \alpha, \nu \circ \sigma)$ such that $i,j<n$, $K \in LIT_\Sigma$, $S \vdash_\Sigma C_i$, $S \vdash_\Sigma C_j$, $\sigma \in SUB_\Sigma$, $\sigma L \underset{\alpha}{\sim} \sigma K$, $\alpha(\sigma L) = \sigma q$ and $\alpha(\sigma K) = \sigma r$.

Since $C_n \subset v\sigma(C_i \cup C_j \cup \{K\})$, by Lemma 9.1 there exists some $\lambda \in SUB_\Sigma$ such that $\theta C_n = \lambda C_n$ and $\lambda v\sigma(C_i \cup C_j \cup \{K\}) \in \mathcal{L}_{\Sigma gr}$.

Therefore $\lambda v\sigma C_i, \lambda v\sigma C_j \in \mathcal{L}_{\Sigma gr}$ and we obtain from the induction hypothesis that $M \cap \lambda v\sigma C_i \neq \emptyset$ and $M \cap \lambda v\sigma C_j \neq \emptyset$.

Case (iii.1) $\lambda v\sigma C \notin M$ or $\lambda v\sigma E(qr) \notin M$. Then by the same argument as in case (ii.1) $M \cap \theta C_n \neq \emptyset$. ◪

Case (iii.2) $\lambda v\sigma L \in M$ and $\lambda v\sigma E(qr) \in M$. Then by Lemma 3.3 (4) $\lambda v\sigma L \underset{\alpha}{\sim} \lambda v\sigma K$ and by Lemma 3.3 (5) $\alpha(\lambda v\sigma L) = \lambda v\sigma q$ and $\alpha(\lambda v\sigma K) = \lambda v\sigma r$. Hence $\lambda v\sigma L \underset{\alpha M}{\rightarrow} \lambda v\sigma K$ and therefore $\lambda v\sigma K \in M$ because M is E-closed.

But then $M \cap \{\lambda v\sigma K\} \neq \emptyset$ and with $\lambda v\sigma K \in \theta C_n$ we obtain that $M \cap \theta C_n \neq \emptyset$. ◪
◪ ◪ ⊠

The soundness of the ΣRP-calculus is now an immediate consequence:

<u>Theorem 9.3</u> (Soundness Theorem for ΣRP)
Let $S \subset \mathcal{L}_\Sigma$ such that $S \vdash_\Sigma \square$. Then S is ΣE-unsatisfiable.

Proof From $S \vdash_\Sigma \square$ and $M \cap \square = \emptyset$ we infer by the Soundness Lemma (9.2) that no E-interpretation M Σ-satisfies S, i.e. S is ΣE-unsatisfiable. ⊠

10 The Sort Theorem

So far we have proved the completeness and the soundness of a resolution and paramodulation based many-sorted calculus, i.e. we have shown that the deductive notions correspond exactly with the notions for truth given for the many-sorted clause language.

But as a necessary prerequisite to judge the usefulness of this calculus for first-order reasoning and for automated theorem proving, we need a comparison with first-order logic and the RP-calculus in particular. In this chapter we establish the connection between the RP- and the ΣRP-calculus: we prove with the **Sort-Theorem** that the semantic and syntactic notions of both calculi are equivalent.

To enable a comparison between both calculi, we have to make explicit the information encoded in the signature and the subsort order. To this effect we will use here **sort symbols** also as **unary predicate symbols**. We assume that $s \in \mathcal{P}_S$ for each $s \in \mathcal{S}$ and define $LIT^{\mathcal{S}}$ and $LIT^{\mathcal{S}}_{\Sigma}$ as the set of all (Σ-)literals in the **extended languages** $\mathcal{L}^{\mathcal{S}}$ and $\mathcal{L}^{\mathcal{S}}_{\Sigma}$. An atom or a literal whose predicate letter is a sort symbol will be called a **sort atom** or a **sort literal** respectively.

Now the information encoded in the signature and in the subsort order relation is represented by certain clauses containing only sort literals and by certain additional sort literals in the given well-sorted clauses:

Suppose we have a term $t \in T_{\Sigma}$ and a sort symbol $s \in \mathcal{S}$ such that $[t] \leq_{\mathcal{S}} s$. Then the intended meaning of the sort literal s(t) is that "t has sort s". Hence s(c) is an unsorted formulation of

$c \in \mathcal{C}_s$. For each non-constant function symbol $f \in \mathcal{F}_{s1...sk,s}$ the rangesort of a term $f(t_1...t_k)$ is s if the arguments t_i agree with the sorts stipulated as domainsorts for f in the signature. This relative sort assignment is expressed formally (in clause notation) as $\{\neg s_1(x_1),...,\neg s_k(x_k), s(f(x_1...x_k))\}$. Such a clause corresponds to a statement like, for instance, " $\forall$ x,y integer(x) $\wedge$ rational(y) $\Rightarrow$ real(x^y) ".

Given $t \in T_\Sigma$ and $s_1, s_2 \in \mathcal{S}$ such that $[t] \leq_{\mathcal{S}} s_1$ and $s_1 <_{\mathcal{S}} s_2$ we obviously want to have that "t has sort s_2" also is a true statement. Given, for instance, that 0 is an integer (i.e. has sort integer$\in\mathcal{S}$) and that integer$<_{\mathcal{S}}$real, it makes sense to say that 0 also is a real number, i.e. real(0). Hence we use $\{\neg s_1(x_1), s_2(x_2)\}$ as a clausal (and unsorted) representation for the subsort order relation. Such a clause corresponds to a statement like " $\forall$ x integer(x) $\Rightarrow$ real(x) ".

The set of the above clauses constitutes the set of **sort axioms**, i.e. a set of clauses representing the signature for function symbols and the subsort order in the unsorted clause language $\mathcal{L}^{\mathcal{S}}$.

It remains to express the sorts of the variables in a given well-sorted clause C: to this effect we simply extend C by all sort literals of form $\neg s(x)$, where $x \in \mathcal{V}(C) \cap \mathcal{V}_s$. The resulting clause is called the **relativization** of the given one, because its truth is relative to the truth of the sort assignments for the variables. The relativization of a clause corresponds to the translation of a statement like " $\forall$ x:integer P(x) " to " $\forall$ x integer(x) $\Rightarrow$ P(x) ".

To define sort axioms formally, we use the **covering relation** $<<_{\mathcal{S}}$ induced by a given sort hierarchy $(\mathcal{S}, <_{\mathcal{S}})$: for a pair of sort symbols s_1 and s_2 we say that s_2 **covers** s_1, in symbols $s_1 <<_{\mathcal{S}} s_2$, iff $s_1 <_{\mathcal{S}} s_2$ and $s_1 <_{\mathcal{S}} s <_{\mathcal{S}} s_2$ for no $s \in \mathcal{S}$.

Definition 10.1

For an $\mathcal{S}$-sorted signature Σ, we define the set A^Σ of all sort axioms of Σ as a minimal subset of $\mathcal{L}^{\mathcal{S}}_\Sigma$ which satisfies

(1) $\{s(c)\}\in A^\Sigma$, if $c\in\mathcal{C}_s$,

(2) $\{\neg s_1(x_1),\ldots,\neg s_k(x_k),s(f(x_1\ldots x_k))\}\in A^\Sigma$, if $x_i\in\mathcal{V}_{s_i}$, $f\in\mathcal{F}_{s_1\ldots s_k,s}$ and $x_i\neq x_j$ for $i\neq j$, and

(3) $\{\neg s_1(y),\ s_2(y)\}\in A^\Sigma$, if $y\in\mathcal{V}_{s_1}$ and $s_1<<_{\mathcal{S}}s_2$. ⊠

Note that A^Σ is finite if $\mathcal{F}$ and $\mathcal{S}$ are finite, because A^Σ is a minimal subset of $\mathcal{L}^{\mathcal{S}}_\Sigma$ satisfying (1) - (3).

Definition 10.2

For a Σ-clause C, the relativization $\hat{C}^\Sigma$ of clause C is a clause in $\mathcal{L}^{\mathcal{S}}_\Sigma$ defined as

$$\hat{C}^\Sigma=\{\neg s_1(x_1),\ldots,\neg s_n(x_n)\}\cup C$$

where $x_i\in\mathcal{V}_{s_i}$ and $\{x_1,\ldots,x_n\}=\mathcal{V}(C)$. The relativization of a Σ-clause set S, denoted $\hat{S}^\Sigma$, in the subset of $\mathcal{L}^{\mathcal{S}}_\Sigma$ defined as

$$\hat{S}^\Sigma=\{\hat{C}^\Sigma\in\mathcal{L}^{\mathcal{S}}_\Sigma \mid C\in S\}.\ ⊠$$

If Σ is known from the context we write $\hat{C}$ instead of $\hat{C}^\Sigma$ and $\hat{S}$ for $\hat{S}^\Sigma$.

Since $\mathcal{L}^{\mathcal{S}}_\Sigma\subset\mathcal{L}^{\mathcal{S}}$, we now have a means to express all sortal information in unsorted predicate logic, i.e. in the unsorted clause language $\mathcal{L}^{\mathcal{S}}$. Hence we are able to relate the semantic and the syntactic notions of both calculi with the Sort-Theorem:

<u>Sort-Theorem for the ΣRP-Calculus</u>

If $(\mathcal{S},<_{\mathcal{S}})$ is a sort-hierarchy, Σ is an $\mathcal{S}$-sorted signature and $S\subset\mathcal{L}_{\Sigma}$, then

(i) S is ΣE-unsatisfiable iff $(\hat{S}\cup A^{\Sigma})$ is E-unsatisfiable, and

(ii) $S^{E}\vdash_{\Sigma}\square$ iff $(\hat{S}^{E}\cup A^{\Sigma})\vdash\square$.

The model-theoretic part (i) of this theorem relates the semantic notions of both calculi, the proof-theoretic part (ii) compares the deductive properties of the RP- and the ΣRP-calculus.

Of course, we have to prove only one part of the Sort-Theorem directly because the other part follows immediately with the soundness and completeness results for both calculi. However, proving the proof-theoretic part directly entails finding a constructive way to translate an (unsorted) refutation in the RP-calculus to a (many-sorted) one in the ΣRP-calculus. To avoid this problem we will prove the model-theoretic part of the Sort-Theorem directly instead. The following diagram relates the deductive and semantic notions of both calculi:

S is ΣE-unsatisfiable $\longleftrightarrow$ $S^{E}\vdash_{\Sigma}\square$

(i) $\updownarrow$ $\updownarrow$ (ii)

$\hat{S}\cup A^{\Sigma}$ is E-unsatisfiable $\longleftrightarrow$ $\hat{S}^{E}\cup A^{\Sigma}\vdash\square$

Figure 10.1 The Sort-Theorem for the ΣRP-calculus

We show both directions of the equivalence (i) in Figure 10.1 independently. We start with the "easy" direction and prove that each E-model of $(\hat{S} \cup A^{\Sigma})$ is also a ΣE-model of S. To this effect we firstly prove that A^{Σ} in fact represents the information encoded in the subsort order and in the signature for function symbols: since we want s(q) to be a true statement for any $q \in T_{\Sigma gr}$ and each $s \in \mathcal{S}$ satisfying $[q] \leq_{\mathcal{S}} s$, we have to prove that each model of A^{Σ} contains s(q):

<u>Lemma 10.1</u>

If M is a model of A^{Σ}, $t \in T_{gr}$ and $s_1, s \in \mathcal{S}$ such that $s_1 \leq s$ and $s_1(t) \in M$, then $s(t) \in M$.

Proof Given the hypotheses of this lemma, we prove by Noetherian induction on s that $s(t) \in M$.

Base Case s is minimal in $(\mathcal{S}, <)$: Then $s_1 = s$ and $s(t) \in M$ by the hypothesis. ∅

Induction Step s is not minimal in $(\mathcal{S}, <)$: Our induction hypothesis is to assume that the lemma holds for each $s_i \in \mathcal{S}$ such that $s_i \ll s$. If $s_1 = s$, then $s(t) \in M$ by the same argument as in the base case and the case is proved.

So let us assume that $s_1 < s$. Then $s_1 \leq s_i$ for some $s_i \in \mathcal{S}$ with $s_i \ll s$ and $\{\neg s_i(x), s(x)\} \in A^{\Sigma}$ by Definition 10.1, hence $\{\neg s_i(t), s(t)\} \in A^{\Sigma}_{gr}$.

By the induction hypothesis we obtain $s_i(t) \in M$, i.e. $\neg s_i(t) \notin M$ because M is an interpretation. But then $s(t) \in M$, because M satisfies A^{Σ}. ∅ ⊠

<u>Definition 10.3</u>

The **kernel** M^{Σ} of A^{Σ} is defined as $M^{\Sigma} = \{s(q) \in LIT^{\mathcal{S}}_{\Sigma gr} \mid s \in \mathcal{S}\}$. ⊠

The kernel M^Σ is an interpretation, because it contains only positive literals. We show that M^Σ is contained in every model of A^Σ, which means that A^Σ represents all information given by the subsort order and by the signature for function symbols:

<u>Lemma 10.2</u>
If M is a model of A^Σ, then $M^\Sigma \subset M$.

Proof Let M be a model of A^Σ and $q \in T_{\Sigma gr}$. We show by structural induction on q that $s(q) \in M$ for each $s \in \mathcal{S}$ satisfying $[q] \leq s$.

Base Case $q \in \mathcal{C}_{s_1}$ with $s_1 \leq s$: By Definition 10.1, $\{s_1(q)\} \in A^\Sigma$, hence $s_1(q) \in M$ because M satisfies A^Σ and with $s_1 \leq s$ we obtain $s(q) \in M$ by Lemma 10.1. ⊘

Induction Step $q \notin \mathcal{C}$: Let $q = f(q_1 \ldots q_k)$, $f \in \mathcal{F}_{s_1 \ldots s_k, s_{k+1}}$, $s_{k+1} \leq s$, $q_i \in T_{\Sigma gr}$ and $[q_i] \leq s_i$ for each i with $1 \leq i \leq k$. Our induction hypothesis is to assume that $s_i(q_i) \in M$ for each i with $1 \leq i \leq k$. By definition 10.1 we know that $\{\neg s_1(x_1), \ldots, \neg s_k(x_k), s_{k+1}(f(x_1 \ldots x_k))\} \in A^\Sigma$, hence $\{\neg s_1(q_1), \ldots, \neg s_k(q_k), s_{k+1}(f(q_1 \ldots q_k))\} \in A^\Sigma_{gr}$.

By the induction hypothesis we obtain that $\neg s_i(q_i) \notin M$, hence $s_{k+1}(f(q_1 \ldots q_k)) = s_{k+1}(q) \in M$ because M satisfies A^Σ. With $s_{k+1} \leq s$ we finally infer by Lemma 10.1 that $s(q) \in M$. ⊘ ⊠

Using the above lemma, we can prove one direction of the equivalence stated in the model-theoretic part of the Sort-Theorem:

<u>Lemma 10.3</u>
If $S \subset \mathcal{L}_\Sigma$ and M is an E-model of $(\hat{S} \cup A^\Sigma)$, then M ΣE-satisfies S.

Proof Let $C \in S$ and $\theta \in SUB_\Sigma$ such that $\theta C \in S_{\Sigma gr}$. Then

$\theta\hat{C} = (\{\neg s_1(q_1),\ldots,\neg s_n(q_n)\} \cup \theta C) \in \hat{S}_{gr}$,

where $q_i \in T_{\Sigma gr}$, $s_i \in \mathcal{S}$ and $[q_i] \leq s_i$ for each i with $1 \leq i \leq n$.

Since M is an E-model of $(\hat{S} \cup A^{\Sigma})$, M also is a model of A^{Σ} and by Lemma 10.2 $M^{\Sigma} \subset M$. Hence by Definition 10.3 $s_i(q_i) \in M$, i.e. $\neg s_i(q_i) \notin M$ because M is an interpretation, and therefore $M \cap \theta\hat{C} = M \cap \theta C$. Since $\theta\hat{C} \in \hat{S}_{gr}$ we know that M satisfies $\theta\hat{C}$, hence $M \cap \theta C \neq \emptyset$ and therefore M Σ-satisfies C. ⊠

The other direction of the equivalence in the model-theoretic part of the Sort-Theorem is obtained with more difficulty. The reason is that a ΣE-model M of S does not necessaryly contain sort literals and therefore M can not satisfy $(\hat{S} \cup A^{\Sigma})$. We will overcome this problem by developing an interpretation M* from M such that

(1) M* interprets each sort literal,
(2) M* is an E-interpretation, and
(3) M* satisfies $(\hat{S} \cup A^{\Sigma})$.

To construct M* from M, we extend M by the so-called M-kernel to an E-interpretation M' which Σ-satisfies $(\hat{S} \cup A^{\Sigma})$. In a second step, M' is extended by the so-called $\mathcal{S}$-complement yielding an E-interpretation M* in which each sort literal L is false, i.e. $L^c \in M^*$, if L is not true in M', i.e. $L \notin M'$. Since we can prove that M* satisfies $(\hat{S} \cup A^{\Sigma})$, M* has all desired properties. The model extensions are formally defined by:

<u>Definition 10.4</u>
For an E-interpretation I, the I-kernel $\ulcorner I \urcorner$ of I is defined as

$\ulcorner I \urcorner = \{s(t) \in LIT^{\mathcal{S}}_{gr} \mid s(q) \in M^{\Sigma}$ and $E(qt) \in I$ for some $q \in T_{\Sigma gr}\}$. ⊠

Definition 10.5

For an E-interpretation I, the $\mathcal{S}$-complement of I, denoted I^C, is defined as

$$I^C=\{\neg s(t)\in LIT_{gr}^{\mathcal{S}} \mid s(t)\notin I \text{ and } s\in\mathcal{S}\}. \boxtimes$$

The following two lemmata prove, that the extension of an E-interpretation I by the I-kernel or by the $\mathcal{S}$-complement of I always yields an E-interpretation.

Lemma 10.4

If $I\subset LIT_{gr}$ is an E-interpretation, then $(I\cup\ulcorner I\urcorner)$ is an E-interpretation.

Proof $(I\cup\ulcorner I\urcorner)$ is an interpretation: We prove that $L^C\notin(I\cup\ulcorner I\urcorner)$ for each $L\in(I\cup\ulcorner I\urcorner)$.

Case (i) $L\in I$: Then $L^C\notin I$ because I is an interpretation. $L^C\notin\ulcorner I\urcorner$ since $\ulcorner I\urcorner$ contains only sort atoms, but L^C cannot be a sort atom because $L\in LIT_{gr}$ by the hypothesis. Hence $L^C\notin(I\cup\ulcorner I\urcorner)$. $\varnothing$

Case (ii) $L\in\ulcorner I\urcorner$: Then $L^C\notin\ulcorner I\urcorner$ because $\ulcorner I\urcorner$ contains only positive literals. $L^C\notin I$ because L^C is a sort literal and $I\subset LIT_{gr}$ by the hypothesis. Hence $L^C\notin(I\cup\ulcorner I\urcorner)$. $\varnothing$ $\varnothing$

$I\cup\ulcorner I\urcorner$ is reflexive: Obvious, because I is reflexive. $\varnothing$

$I\cup\ulcorner I\urcorner$ is E-closed: Let $L\in(I\cup\ulcorner I\urcorner)$, $K\in LIT_{gr}^{\mathcal{S}}$ and $\alpha\in SEL^+$ such that $L\underset{\alpha}{\rightarrow}_{(I\cup\ulcorner I\urcorner)}K$. Then $E(\alpha(L)\alpha(K))\in(I\cup\ulcorner I\urcorner)$, and since $\ulcorner I\urcorner$ contains only sort atoms we obtain $E(\alpha(L)\alpha(K))\in I$, i.e. $L\underset{\alpha}{\rightarrow}_I K$.

If $L\in I$, then $K\in I$ by the E-closure of I and we are finished. So let us assume that $L\in\ulcorner I\urcorner$, i.e. by Definition 10.4 there exist $s\in\mathcal{S}$, $t\in T_{gr}$ and $q\in T_{\Sigma gr}$ such that $L=s(t)$, $s(q)\in M^{\Sigma}$ and $E(qt)\in I$.

From $L \underset{\alpha I}{\rightarrow} K$ we obtain $L \underset{\alpha}{\sim} K$, hence $K=s(r)$ for some $r \in T_{gr}$. Let $\alpha_1, \alpha_2 \in SEL$ such that $\alpha_1(s(t))=t$ and $\alpha_2(E(qt))=t$. Then $\alpha=\beta \circ \alpha_1$ for some $\beta \in SEL^*$ and with $L \underset{\alpha I}{\rightarrow} K$ we obtain by Lemma 3.3 (3) $t \underset{\beta I}{\rightarrow} r$. Hence $\alpha_2(E(qt)) \underset{\beta I}{\rightarrow} \alpha_2(E(qr))$ and with $E(qt) \underset{\alpha_2}{\sim} E(qr)$ we infer by Lemma 3.3 (3) $E(qt) \underset{\beta \circ \alpha_2 I}{\rightarrow} E(qr)$.

Hence $E(qr) \in I$, because $E(qt) \in I$ and I is E-closed. Using $s(q) \in M^{\Sigma}$, by Definition 10.4 we finally obtain $s(r)=K \in \ulcorner I \urcorner$. ☒ ⊠

<u>Lemma 10.5</u>
If $I \subset LIT_{gr}^{\mathcal{S}}$ is an E-interpretation, then $(I \cup I^C)$ is an E-interpretation.

Proof $(I \cup I^C)$ is an interpretation: Let $L \in (I \cup I^C)$. If $L \in I$, then $L^C \notin I$ because I is an interpretation, and $L^C \notin I^C$ by Definition 10.5. If $L \in I^C$, then $L^C \notin I$ by definition, and $L^C \notin I^C$ because I^C contains only negative literals. Hence in either case $L^C \notin (I \cup I^C)$. ☒

$(I \cup I^C)$ is reflexive: Obvious, because I is reflexive. ☒

$(I \cup I^C)$ is E-closed: Let $L \in (I \cup I^C)$, $K \in LIT_{gr}^{\mathcal{S}}$ and $\alpha \in SEL^+$ such that $L \underset{\alpha (I \cup I^C)}{\rightarrow} K$. Then $E(\alpha(L)\alpha(K)) \in (I \cup I^C)$, and since I^C contains only negative sort literals we obtain $E(\alpha(L)\alpha(K)) \in I$, i.e. $L \underset{\alpha I}{\rightarrow} K$.

If $L \in I$, then $K \in I$ by the E-closure of I and we are finished. So let us assume that $L \in I^C$. Suppose that $K^C \in I$. From $L \underset{\alpha I}{\rightarrow} K$ we obtain $L^C \underset{\alpha I}{\rightarrow} K^C$, hence by Lemma 3.4 (1) $K^C \underset{\alpha I}{\rightarrow} L^C$, i.e. $L^C \in I$ because I is E-closed. But then $L \notin I^C$ by Definition 10.5. ∇ Hence $K^C \notin I$, i.e. $K \in I^C$. ☒ ⊠

With the above lemma, it is obvious that $(I \cup I^C)$ is an E-interpretation which interprets **each** sort literal, i.e. $s(t) \in (I \cup I^C)$ or $\neg s(t) \in (I \cup I^C)$ for all $s \in \mathcal{S}$ and all $t \in T_{gr}$, provided I is an E-interpretation.

Having proved that the extension of E-interpretations I by the I-kernel and the $\mathcal{S}$-complement of I again yields an E-interpretation, it remains to prove that we obtain an E-model of $(\hat{S}UA^{\Sigma})$ from a ΣE-model of S using the above extensions:

Lemma 10.6

If $S \subset \mathcal{L}_{\Sigma}$ is ΣE-satisfiable, then $(\hat{S}UA^{\Sigma})$ is E-satisfiable.

Proof Let M be a minimal ΣE-model of S. Then M is an E-model of $S_{\Sigma gr} \subset \mathcal{L}_{\Sigma gr}$, and since M is minimal we have $M \subset LIT_{gr}$. Hence by Lemma 10.4 $(M \cup \ulcorner M \urcorner)$ is an E-interpretation and using Lemma 10.5 we obtain that

$$M^* = (M \cup \ulcorner M \urcorner) \cup (M \cup \ulcorner M \urcorner)^C$$

also is an E-interpretation. Let $C \in (\hat{S}UA^{\Sigma})$, i.e.

$$C = \{\neg s_1(x_1), \ldots, \neg s_n(x_n)\} \cup D$$

where $D \in \mathcal{L}_{\Sigma}^{\mathcal{S}}$ and no literal in D is a negative sort literal, $\mathcal{V}(D) = \{x_1, \ldots, x_n\}$ and $[x_i] = s_i$ for each i with $1 \leq i \leq n$, cf. Definitions 10.1 and 10.2. Let $\theta \in SUB_{gr}$ such that $\theta = \{x_1 \leftarrow t_1, \ldots, x_n \leftarrow t_n\}$, i.e.

$$\theta C = (\{\neg s_1(t_1), \ldots, \neg s_n(t_n)\} \cup \theta D) \in (\hat{S}UA^{\Sigma})_{gr}.$$

We prove that M* satisfies θC.

Case (i) $s_i(t_i) \notin \ulcorner M \urcorner$ for some i with $1 \leq i \leq n$: Then $s_i(t_i) \notin (M \cup \ulcorner M \urcorner)$ because M contains no sort literal, hence $\neg s_i(t_i) \in (M \cup \ulcorner M \urcorner)^C$ by Definition 10.5. But then $\neg s_i(t_i) \in M^* \cap \theta C$, i.e. M* satisfies θC.$\varnothing$

Case (ii) $s_i(t_i) \in \ulcorner M \urcorner$ for each i with $1 \leq i \leq n$: Then by Definition 10.4 there exist $q_1, \ldots, q_n \in T_{\Sigma gr}$ such that

$$s_i(q_i)\in M^{\Sigma} \text{ and } E(q_i t_i)\in M$$

for each i with $1\leq i\leq n$. Let $\lambda\in SUB$ such that $\lambda=\{x_1\leftarrow q_1,\ldots,x_n\leftarrow q_n\}$. From $s_i(q_i)\in M^{\Sigma}$ we infer $\lambda\in SUB_{\Sigma}$.

We prove that $(M\cup\ulcorner M\urcorner)\cap\lambda D\neq\emptyset$.

Case (ii.1) $s(r)\in D$ for some $s\in\mathcal{S}$ and some $r\in T$. With $D\in\mathcal{L}^{\mathcal{S}}_{\Sigma}$ we obtain $s(\lambda r)\in LIT^{\mathcal{S}}_{\Sigma gr}$. Then by Definition 10.3 $s(\lambda r)\in M^{\Sigma}$, hence by Definition 10.4 $s(\lambda r)\in\ulcorner M\urcorner$ because $E(\lambda r\lambda r)\in M$, i.e. $s(\lambda r)\in(M\cup\ulcorner M\urcorner)\cap\lambda D$. ▨

Case (ii.2) $s(r)\notin D$ for each $s\in\mathcal{S}$ and each $r\in T$: Then $C=\hat{D}$, i.e. $D\in S$ and $\lambda D\in S_{\Sigma gr}$. Since M satisfies $S_{\Sigma gr}$ we know that $M\cap\lambda D\neq\emptyset$, i.e. $(M\cup\ulcorner M\urcorner)\cap\lambda D\neq\emptyset$. ▨

Hence there exists some $L\in D$ such that $\lambda L\in(M\cup\ulcorner M\urcorner)$. Let $\{\alpha_1,\ldots,\alpha_h\}=\{\alpha\in SEL^{+}\mid\alpha(L)=x_i$ for some $x_i\in\mathcal{V}(D)\}$ and let $K_1,\ldots,K_{h+1}\in LIT^{\mathcal{S}}_{gr}$ such that $K_1=\lambda L$, $K_j\underset{\alpha_j}{\sim}K_{j+1}$, $\alpha_j(K_j)=\lambda\alpha_j(L)$ and $\alpha_j(K_{j+1})=\theta\alpha_j(L)$ for all $j\in\{1,\ldots h\}$.

Then $K_{h+1}=\theta L$ and $K_1\ \vec{\alpha}_1(M\cup\ulcorner M\urcorner)\ K_2\ldots K_h\ \vec{\alpha}_h(M\cup\ulcorner M\urcorner)\ K_{h+1}$, i.e.

$$K_1 \overset{*}{\rightarrow}_{\mathcal{R}(M\cup\ulcorner M\urcorner)} K_{h+1}$$

because for each j there is some i such that $\alpha_j(K_j)=\lambda x_i=q_i$, $\alpha_j(K_{j+1})=\theta x_i=t_i$ and $E(q_i t_i)\in M$.

With $K_1=\lambda L\in(M\cup\ulcorner M\urcorner)$ we obtain $K_{h+1}=\theta L\in(M\cup\ulcorner M\urcorner)$ because M* is E-closed. Hence $\theta L\in(M\cup\ulcorner M\urcorner)\cap\theta D$, i.e. M* satisfies θC. ▨ ▨ ⊠

Using the above results, the Sort-Theorem is now an immediate consequence:

<u>Theorem 10.7</u> (Sort-Theorem for the ΣRP-Calculus)
If $(\mathcal{S},<_{\mathcal{S}})$ is a sort-hierarchy, Σ is an $\mathcal{S}$-sorted signature and $S\subset\mathcal{L}_{\Sigma}$, then

(i) S is ΣE-unsatisfiable iff $(\hat{S}\cup A^{\Sigma})$ is E-unsatisfiable, and

(ii) $S^{E}\vdash_{\Sigma}\square$ iff $(\hat{S}^{E}\cup A^{\Sigma})\vdash\square$.

Proof (i) If S is ΣE-unsatisfiable, then by Lemma 10.3 $(\hat{S}\cup A^{\Sigma})$ is E-unsatisfiable and if $(\hat{S}\cup A^{\Sigma})$ is E-unsatisfiable, then S is ΣE-unsatisfiable by Lemma 10.6. ⌀

(ii) $S^{E}\vdash_{\Sigma}\square$ iff S^{E} is ΣE-unsatisfiable because the ΣRP-calculus is sound and complete (Theorems 9.3 and 8.4). S^{E} is ΣE-unsatisfiable iff $(\hat{S}^{E}\cup A^{\Sigma})$ is E-unsatisfiable by (i). $(\hat{S}^{E}\cup A^{\Sigma})$ is E-unsatisfiable iff $(\hat{S}^{E}\cup A^{\Sigma})\vdash\square$, because the RP-calculus is sound and complete [Robinson 1965, Wos and Robinson 1973]. ⌀ ⊠

11 An Automated Theorem Prover for the ΣRP-Calculus

In this chapter a brief survey is presented of how an automated theorem prover ATP based on the RP-calculus can be modified to obtain an automated theorem prover ATP_Σ for the ΣRP-calculus. The necessary modifications of the ATP concern

- the first-order input language,
- the input language compiler,
- the skolemization routine,
- the unification algorithm, and
- the computation of resolvents, factors and paramodulants.

For the remainder of this chapter we assume familiarity with the notions and definitions given in Chapters 3 and 4. For notions from other chapters explicit references will be provided.

The Input Language First of all, we have to define a many-sorted first-order notation L_Σ, which provides language constructs to define the set of sort symbols $\mathcal{S}$, a subsort order relation $<_{\mathcal{S}} \subset \mathcal{S} \times \mathcal{S}$ and some $\mathcal{S}$-sorted signature Σ. Since the ΣRP-calculus is a true generalization of the RP-calculus, it is useful to define L_Σ such that L_Σ is a superset of the first-order input language L used for the unsorted ATP, cf. [Walther 1982]. This allows unsorted formulas to be proved with the many-sorted prover ATP_Σ without translating them into an explicit many-sorted notation (where $\mathcal{S}$ is a singleton of course).

In the unsorted input language L of the MKRP system [Bläsius et al. 1981, Eisinger and Ohlbach 1986], for instance, we use all x ... and ex x ... for universal and for existential quantification respectively. The arities of function and predicate symbols are determined by the number of arguments in the first usage of the respective symbol in a sequence of formulas.

For the many-sorted language we need a means to define an $\mathcal{S}$-sorted signature. We use all x:s ... and ex x:s ... to denote that the variable x in a quantification has sort s, i.e. $x \in \mathcal{V}_s$. The signature for function and predicate symbols is defined by means of the so-called type declaration:

type c:s defines $c \in \mathcal{C}_s$,

type f(s_1 ... s_n):s defines $f \in \mathcal{F}_{s_1 \ldots s_n, s}$,and

type p(s_1 ... s_n) defines $p \in \mathcal{P}_{s_1 \ldots s_n}$.

Since we are interested in equality reasoning, we assume that each sort hierarchy $(\mathcal{S}, <_{\mathcal{S}})$ is a tree structure (where we use the identifier any instead of the root sort symbol $s\Delta$, cf. Chapter 4). Formulas of the unsorted language LCL_{Σ} are treated in the following way: quantifications like all x... and ex x... are implicitly treated as all x:any ... and ex x:any ... respectively. Like for L, the arities of function and predicate symbols are determined by the number of arguments in the first usage of the respective symbols in a list of formulas. If the first usage is not in a type declaration, we define $f \in \mathcal{F}_{any \ldots any, any}$ and $p \in \mathcal{P}_{any \ldots any}$ for a function or predicate symbol respectively.

The subsort order relation is defined by means of the so-called subsort declaration: starting with $<_{\mathcal{S}} = \emptyset$,

sort $s_1, \ldots, s_m : t_1, \ldots, t_n$

extends $<_{\mathcal{S}}$ by $s_i <<_{\mathcal{S}} t_j$ for all i and j with $1 \leq i \leq m$ and $1 \leq j \leq n$ (where $<<_{\mathcal{S}}$ is the covering relation, cf. Chapter 10).

Finally, we need a means to define the set of sort symbols $\mathcal{S}$: starting with $\mathcal{S}=\{any\}$, $\mathcal{S}$ is dynamically extented by each new sort symbol used in a quantification, in a type declaration or in a subsort declaration.

The Compiler The compiler of a theorem prover tests whether a given input string satisfies the rules of syntax and those of the static semantics (i.e. that function symbols are used with a proper arity etc.) of a first-order language L with the usual junctors, universal and existential quantifiers. It produces as 'code' a first-order formula in a certain representation which is appropriate for further processing.

The rules of the static semantics have to be extended such that only formulas from the set of all well-sorted first-order formulas $L_{\Sigma} \supset L$ will be accepted: for each atomar formula A in a formula given as input, the compiler has to determine whether A is a well-sorted atomar formula, i.e.

$$[\alpha(A)] \leq_{\mathcal{S}} [A]_{\alpha} \text{ for each } \alpha \in SEL^{+} \text{ such that } \alpha(A)\downarrow.$$

This problem is the same as for programming languages with sorts (often called types), e.g. PASCAL or ADA, and hence can be solved using the well known techniques of compiler construction.

In addition the compiler has to perform a 'semantic' test to check whether $<_{\mathcal{S}}$ defined by the subsort declarations is in fact an order relation.

The Skolemization Routine On skolemization of a first-order formula (given in a certain format) each occurence of an existentially quantified variable symbol y in an atomar formula is replaced by a skolem-term t, and all existential quantifiers are removed.

The skolem-term t consists of a new function symbol f followed by a (possibly empty) sequence $x_1,\ldots,x_n$ of variable symbols as arguments, where each x_i is a universally quantified variable symbol and the variable symbol y, which was replaced by t, is in the scope of exactly the universal quantifiers for the variable symbols x_i.

For Σ-skolemization, i.e. skolemization under sorts, this process is the same for each formula in L_Σ. Also the signature Σ has to be extended, yielding a signature Σ* for all the new function symbols introduced by the Σ-skolemizations. We define

$$f \in \mathcal{F}_{[x_1]\ldots[x_n],[y]}$$

where f, x_i and y are defined as above, and it is obvious that Σ-skolemization transforms well-sorted formulas (of L_Σ) into well-sorted formulas (of L_{Σ^*}).

To be correct, we have to show that Σ-skolemization maintains Σ-(un)satisfiability: From [Oberschelp 1962] we obtain the semantic notions for L_Σ, and in particular a notion of Σ-unsatisfiability for formulas $\Phi \in L_\Sigma$. Let $L^{\mathcal{G}}$ be the extended language of L, where sort symbols are used as unary predicate symbols (cf. Chapter 10). Then by the Sort-Theorem of [Oberschelp 1962], each formula

$$\Phi \in L_\Sigma \text{ is } \Sigma\text{-unsatisfiable iff } (\{\hat{\Phi}\} \cup A^\Sigma) \subset L^{\mathcal{G}}_\Sigma \text{ is unsatisfiable.}$$

Consider the following diagram:

$$\begin{array}{ccc} \Phi \in L_{\Sigma} & \xrightarrow{(1)} & (\{\hat{\Phi}\} \cup A^{\Sigma}) \subset L_{\Sigma}^{\varphi} \\ \Big\downarrow (3) & & \Big\downarrow (2) \\ \tilde{\Phi} \in L_{\Sigma^*} & \xrightarrow[(4)]{} & (\{\hat{\tilde{\Phi}}\} \cup A^{\Sigma^*}) \underset{(5)}{\equiv} (\{\tilde{\hat{\Phi}}\} \cup A^{\Sigma}) \subset L_{\Sigma^*}^{\varphi} \end{array}$$

Figure 11.1 Skolemization and Σ-Skolemization

Here $\tilde{\Phi}$ and $\tilde{\hat{\Phi}}$ denote the formulas which are obtained from the formula Φ by (Σ-)skolemization on (3) and (2). On (1) and (4) a formula is replaced by its relativization and the set of sort axioms for the given signatures Σ and Σ^* respectively (cf. Chapter 10).

We can prove the equivalence (5) of Figure 11.1 under the assumption that equal variables are replaced by equal skolem-terms by the (Σ-)skolemizations of (2) and (3).

Hence we obtain from Figure 11.1 that each Σ-skolemized formula $\tilde{\Phi} \in L_{\Sigma^*}$ is Σ^*-unsatisfiable iff $\Phi \in L_{\Sigma}$ is Σ-unsatisfiable, because (1) and (4) maintain (un)satisfiability by the Sort-Theorem of [Oberschelp 1962], and (2) is the skolemization in the one-sorted language and hence leaves (un)satisfiability unchanged.

Example 11.1

Let $\mathcal{S}=\{s_1,s_2\}$, $<_{\mathcal{S}}=\emptyset$ and $\mathcal{F}_{w,s}=\emptyset$ for all $w\in\mathcal{S}^*$ and $s\in\mathcal{S}$ and consider

(*) $\Phi=<\text{all } x{:}s_1 \text{ ex } y{:}s_2 \ P(x\ y)> \ \in L_{\Sigma}$.

Then $A^{\Sigma}=\emptyset$, hence

(1) $\{\hat{\Phi}\}\cup A^{\Sigma}=\{<\text{all } x \ s_1(x) \Rightarrow (\text{ex } y \ s_2(y) \wedge P(x\ y))>\}$

and skolemization of (1) yields

(2) $\{\tilde{\hat{\Phi}}\}\cup A^{\Sigma}=\{<\text{all } x \ s_1(x) \Rightarrow (s_2(f(x)) \wedge P(x\ f(x)))>\}$.

Σ-skolemization of the formula (*) yields

(3) $\tilde{\Phi}=<\text{all } x{:}s_1 \ P(x\ f(x))> \ \in L_{\Sigma^*}$, where $f\in\mathcal{F}^*_{s_1,s_2}$,

hence we obtain with relativization and the set of sort axioms (for Σ^*)

(4) $\{\hat{\tilde{\Phi}}\}\cup A^{\Sigma^*}=\{<\text{all } x \ s_1(x) \Rightarrow P(x\ f(x))>\} \ \cup$
$\{<\text{all } z \ s_1(z) \Rightarrow s_2(f(z))>\}$.

Note that the sets of formulas in (2) and (4) are semantically equivalent. ⊠

The Unification Algorithm The unification algorithm (1) decides whether a pair of terms (or atoms) is unifiable and (2) returns a most general unifier for the pair or a failure indication, depending on the result of the former test. To modify a unification algorithm in order to obtain a many-sorted version of it we must

(1) insert additional tests, which check whether the additional constraints for unifiable terms imposed by the subsort order relation are satisfied, and

(2) provide a means to generate auxiliary variables, which are necessary to produce a minimal and complete set of well-sorted unifiers for a pair of Σ-unifiable variables, if non-forest structured sort hierarchies are used.

Let us consider the following algorithm $MCSU_\Sigma$ (abbreviating **Mi**nimal and **C**omplete **S**et of Σ-**Un**ifiers):

```
function MCSU_Σ(t_1,t_2∈T_Σ∪AT_Σ):2^(SUB_Σ) =

(1) if t_1=t_2
      then return({∅}) (* i.e. {ε} *)
    fi

(2) if t_1∈V
      then
   (2.1) if t_1∈V(t_2) (* occur check *)
           then return(∅) (* occur failure *)
         fi

   (2.2) if [t_2]≤_𝒮[t_1]
           then return({{t_1←t_2}})
         fi
```

```
(2.3) if t2∉V or lbs_S{[t1],[t2]}=∅
        then return(∅)
      fi

      assume t2∈V, max_S(lbs_S{[t1],[t2]})={s1,...,sk}, k≥1
(2.4) if [t1]<_S[t2]
        then return({{t2←t1}})
        else for j:=1 (1) k do
                 zj:=newV_sj
             done
(2.5)        return({{t1←z1,t2←z1},...,{t1←zk,t2←zk}})
      fi
 fi

 assume t1∉V
(3) if t2∈V
     then
      if t2∈V(t1) (* occur check *)
        then return(∅) (* occur failure *)
      fi


(3.1) if [t1]≤_S[t2]
        then return({{t2←t1}})
        else return(∅)
      fi
 fi


 assume t1=F(q1...qn), t2=G(r1...rm), n≥0, m≥0
(4) if F≠G (*clash check *)
     then return(∅) (* clash failure *)
    fi
```

```
(5) assume t1=f(q1...qn), t2=f(r1...rn), n>0
    U:={Ø} (* i.e. {ε} *)
    i:=0
    repeat
      i:=i+1
      U*:=U
      U:=Ø
      for all σ∈U* do
          assume σ={x1←s1,...,xh←sh} , h≥0
  (5.1)   for all θ∈MCSU_Σ(σqi,σri) do
              U:=U ∪ {θ ∪ {x1←θs1,...,xh←θsh}} (* σ:=θ∘σ *)
          done
      done
    until i=n or U=Ø

    return(U)
 end
```

Figure 11.2 An algorithm for many-sorted unification

We assume that the semantics of the statements in the algorithm are obvious without a precise and formal definition. We only remark, that the execution of the algorithm is aborted after an execution of a return statement, yielding the argument of this statement as result of the algorithm. lbs_{φ} and max_{φ} are defined as in Chapter 7.

Given a pair of well-sorted terms or atoms as input, the algorithm returns a set of well-sorted substitutions. A substitution is represented by a finite set of variable/term pairs, written as x←t, where the empty set Ø represents the empty substitution ε. Hence returning Ø for a given input t_1,t_2 means that t_1 and t_2 are not Σ-unifiable while returning {Ø} means that ε Σ-unifies t_1 and t_2, i.e. $t_1=t_2$.

Let us consider the statements of the many-sorted unification algorithm in detail:

Statements (1) and (2.1) are the usual test for identity and for "occur"-conflicts respectively. Statement (2.2) tests whether the result $\{t_1 \leftarrow t_2\}$ which would be returned by the unsorted algorithm as a most general unifier is a well-sorted substitution. If so, we do not need auxiliary variables for this case. Statement (2.3) tests whether a well-sorted solution to the given unification problem is possible at all (cf. Lemma 7.3): for "$t_2 \notin \mathcal{V}$ or $lbs_{\mathcal{S}}\{[t_1],[t_2]\}=\emptyset$" the unification problem cannot have a well-sorted solution, because there are no common well-sorted instances of t_1 and t_2 in this case.

The then-part of statement (2.4) is symmetric to statement (2.2): again the result returned by the unsorted algorithm as a most general unifier here is also correct for the many-sorted problem. On execution of the else-part, we find that $\{t_1,t_2\} \subset \mathcal{V}$ and $\sqcap[\{t_1,t_2\}] \notin [\{t_1,t_2\}]$. Hence we have to compute all the auxiliary variables as given as in Definition 7.1 in order to obtain a minimal and complete set of well-sorted unifiers for the unification problem in this case. Note that $new\mathcal{V}_s$ generates a fresh variable of sort s, i.e. a variable which was never used before.

Statement (3) is symmetric to statement (2) with t_1 and t_2 interchanged. But since we know here that $t_1 \notin \mathcal{V}$, a test similiar to the test in statement (2.3) would always be affirmative. Hence a failure indication is returned unconditionally if the test of statement (3.1) evaluates to false. Statement (4) looks for "clash"-conflicts as in the unsorted case.

The statements following (5) treat the recursion case of the many-sorted unification algorithm in the same way as in the unsorted algorithm: most general solutions of a unification problem are computed by stepwise combination (using functional composition) of most general solutions of subproblems. The invariant of the repeat-loop following (5) is

$$U = mcU_{\Sigma}(\{q_1, \ldots, q_i, r_1, \ldots, r_i\})$$

We remark that $\{q_1, \ldots, q_i, r_1, \ldots, r_i\} = \emptyset$ for i=0 and $mcU_{\Sigma}(\emptyset) = \{\varepsilon\}$, because ε is a most general unifier of $\emptyset$. Each of the most general solutions $\sigma \in mcU_{\Sigma}(\{q_1, \ldots, q_i, r_1, \ldots, r_i\})$ computed so far (and collected in U) is applied to q_{i+1} and r_{i+1}. A minimal and complete set of well-sorted unifiers is computed for the instantiated terms $\sigma q_{i+1}, \sigma r_{i+1}$ and each of the generated well-sorted unifiers θ is composed with the former solution σ, yielding a new member of $mcU_{\Sigma}(\{q_1, \ldots, q_{i+1}, r_1, \ldots, r_{i+1}\})$. Note that deadend solutions, i.e. substitutions σ which instantiate q_{i+1} and r_{i+1} such that σq_{i+1} and σr_{i+1} are not Σ-unifiable, are eliminated for further processing (i.e. do not contribute to the final result), because the body of the for-loop (5.1) is not executed in this case. The many-sorted unification algorithm terminates with a failure indication (viz. $U=\emptyset$), if all members σ of U prevent further Σ-unifications.

Statement (2.5) together with the nested loops in statements following (5) give rise to a combinatorical explosion. Depending on a particular sort hierarchy (and the given unification problem of course) $MCSU_{\Sigma}$ may compute sets of arbitrarily large (but always finite) cardinality. Hence computing a minimal and complete set of well-sorted unifiers may become very expensive.

However, this can never happen if forest-structured sort hierarchies are used (cf. Theorem 7.16). In this case one of the conditions in the statements (2.2), (2.3) and (2.4) is always satisfied, and therefore statement (2.5) never is executed. As a consequence, the set U in the statements following (5) is at most a singleton, and it can easily be seen that $MCSU_{\Sigma}(t_1, t_2)$ contains exactly one well-sorted most general unifier of t_1 and t_2, provided this pair of terms is Σ-unifiable at all.

A combinatorical explosion in the case of a non-forest sort hierarchy $(\mathcal{S},<_{\mathcal{S}})$ is also impossible, if $(\mathcal{S},<_{\mathcal{S}})$ is a meet-semilattice, cf. Corollary 7.14. Then the parameter k in statement (2.5) is always 1, hence the set U in the statements following (5) (and obviously also $MCSU_{\Sigma}$) is at most a singleton.

We can take advantage of this result by embedding $(\mathcal{S},<_{\mathcal{S}})$ into a meet-semilattice $(\mathcal{S}^{+},<_{\mathcal{S}^{+}})$ (cf. [McNeille 1937]) : starting with $(\mathcal{S},<_{\mathcal{S}})$, the many-sorted theorem prover ATP_{Σ} has to compute a sort hierarchy $(\mathcal{S}^{+},<_{\mathcal{S}^{+}})$ satisfying

(1) $\mathcal{S} \subset \mathcal{S}^{+}$,

(2) for all $s_1,s_2 \in \mathcal{S}$: $s_1 <_{\mathcal{S}} s_2$ iff $s_1 <_{\mathcal{S}^{+}} s_2$, and

(3) for all $s_1,s_2 \in \mathcal{S}^{+}$ at most one $s \in \mathcal{S}^{+}$ exists such that:

$$s \ll_{\mathcal{S}^{+}} s_1 \text{ and } s \ll_{\mathcal{S}^{+}} s_2 .$$

(4) $\mathcal{S}^{+}$ is a minimal set satisfying (1), (2) and (3).

Using the results of Chapter 7, it can be proved that the many-sorted unification algorithm $MCSU_{\Sigma}$ always terminates and that $MCSU_{\Sigma}(t_1,t_2)=mcU_{\Sigma}(\{t_1,t_2\})$ for all $t_1,t_2 \in T_{\Sigma} \cup AT_{\Sigma}$.

Computation of Resolvents, Factors and Paramodulants Based on the definitions presented in Chapter 4, the generalization of the algorithms which compute resolvents, factors and paramodulants to the many-sorted case is straightforward. Nevertheless they are presented here for sake of completeness.

Given a pair of clauses C and D and a pair of literals L and K, there is at most one resolvent of C and D upon L and K in the unsorted case (modulo variable renaming). For the many-sorted case, we have to compute an arbitrarily large (but always finite) set of (independent) Σ-resolvents of C and D upon L and K - one for each member of $mcU_\Sigma(\{|L|,|K|\})$.

```
function RES_Σ(C∈𝓛_Σ,L∈LIT_Σ,D∈𝓛_Σ,K∈LIT_Σ):2^𝓛Σ=
   R:=Ø
   if L∈C and K∈D
     then if L complementary K
            then for all σ∈MCSU_Σ(|L|,|K|) do
                      R:=R U {σ(C-L)Uσ(D-K)}
                 done
          fi
   fi
   return(R)
end
```

Figure 11.3 Computation of many-sorted resolvents

It is easily verified that each member of $RES_\Sigma(C,L,D,K)$ is a Σ-resolvent of C and D upon L and K. Obviously, the many-sorted theorem prover ATP_Σ has to compute all these Σ-resolvents in order to be refutation complete (cf. Lemma 8.1) - at least if ATP_Σ is organized as a breadth-first search procedure without strategies or refinements (as, for instance, "set-of-support", "unit-refutation" etc.). Since the cardinality of $RES_\Sigma(C,L,D,K)$

is at most equal to the cardinality of $MCSU_\Sigma(|L|,|K|)$, the costs of computing RES_Σ depend only on the costs of computing $MCSU_\Sigma$. Hence for a sort hierarchy, which is tree-structured or is a meet-semilattice, the computation of RES_Σ is as cheap (or expensive) - in terms of unification - as the computation of a resolvent in the unsorted theorem prover ATP.

The Σ-factors of a Σ-clause are computed by the following algorithm:

```
function FAK_Σ(C∈ℒ_Σ):2^ℒΣ=
   F:=∅
   D:=C
   for all L∈C do
      D:=D-L
      for all K∈D do
         if not (L complementary K)
            then for all σ∈MCSU_Σ(|L|,|K|) do
                        F:=F ∪ FAK_Σ(σ(C))
                     done
         fi
      done
   done
   return(F ∪ {C})
end
```

Figure 11.4 Computation of many-sorted factors

Here the same comment applies as for the computation of Σ-resolvents: given a pair of clauses C and D such that $D \subset C$, there is at most one factor σC of C upon D (modulo variable renaming) in the unsorted case (where σ is a most general unifier of D). For the many-sorted case there are as many Σ-factors of C upon D, as there are Σ-unifiers in $mcU_\Sigma(D)$. It is easily verified that FAK_Σ computes all Σ-factors of a given well-sorted clause. In order to be refutation complete, the many-sorted theorem prover ATP_Σ has to compute all these Σ-factors (cf. Lemmata 8.1 and 8.2).

Given a pair of clauses C and D, a pair of literals L and $K=E(qr)$ and a selector $\alpha \in SEL^+$, there are at most two paramodulants of C and D upon L and K regarding α in the unsorted case (modulo variable renaming). For the many-sorted case, we have to compute an arbitrarily large (but always finite) set of (independent) Σ-paramodulants of C and D upon L and K - at most one for each member of $mcU_\Sigma(\{\alpha(L),q\}) \cup mcU_\Sigma(\{\alpha(L),r\})$. But in contrast to the cases of resolution and factoring, we also have to compute all the coercion substitutions which are neccessary to obtain well-sorted paramodulants (cf. Chapters 4 and 8).

```
function PAR_Σ(C∈𝓛_Σ,L∈LIT_Σ,D∈𝓛_Σ,K∈LIT_Σ,α∈SEL⁺):2^𝓛_Σ=
   P:=∅
   if L∈C and K∈D
     then if K∈AT^E_Σ
            then assume K=E(qr)
                 let M∈LIT such that L~_α M and α(M)=r
                 for all σ∈MCSU_Σ(α(L),q) do
                    z:=new𝒱_[L]_α

                    for all ν∈MCSU_Σ(z,σr) do
                       P:=P ∪ {νσ(C-L)∪νσ(D-K)∪{νσM}}
                    done
                 done
```

```
            let M∈LIT such that L~_α M and α(M)=q
            for all σ∈MCSU_Σ(α(L),r) do
                z:=newV_[L]_α

                for all ν∈MCSU_Σ(z,σq) do
                    P:=P ∪ {νσ(C-L)∪νσ(D-K)∪{νσM}}
                done
            done
        fi
    fi
    return(P)
end
```

Figure 11.5 Computation of many-sorted paramodulants

It is easily verified that $PAR_\Sigma(C,L,D,E(qr),\alpha)$ contains all Σ-paramodulants of C and D upon L and E(qr) regarding α. Since we compute the coercion substitutions as members of a minimal and complete set of unifiers using $MCSU_\Sigma$, we can make the computation of Σ-paramodulants as cheap as in the unsorted case (in terms of unification) by using sort hierarchies, which are forest-structured or are meet-semilattices. The many-sorted theorem prover ATP_Σ has to compute all these Σ-paramodulants in order to be refutation complete (cf. Lemma 8.2).

After computation of Σ-resolvents, Σ-factors and Σ-paramodulants, each variable symbol in each of these clauses has to be replaced by a fresh one, using appropriate Σ-renaming substitutions.

12 Some Experiences with a Many-Sorted Theorem Prover

In this chapter we shall discuss the practical relevance of the proposed many-sorted calculus by few examples. We compare the solution of some unsorted theorem proving problems with their many-sorted solutions using the Markgraf Karl Refutation Procedure (MKRP), i.e. an automated theorem prover developed at the universities of Karlsruhe and Kaiserslautern [Bläsius et al. 1981, Eisinger and Ohlbach 1986]. The MKRP-system is based on the extended connection graph proof procedure [Kowalski 1975, Eisinger et al. 1978, Siekmann and Wrightson 1980, Walther 1981] and was extended to a many-sorted system by the modifications as described in Chapter 11. The system is implemented in INTERLISP and runs on a SIEMENS 7760 machine.

A starting point for the comparison of both calculi is the Sort-Theorem (10.7). We obtain a shorter refutation from a smaller set of smaller clauses, if we prove $S^E \vdash_\Sigma \square$ instead of $(\hat{S}^E \cup A^\Sigma) \vdash \square$. But it is not fair, in some cases, to compare Σ-refutations of S^E with refutations of $\hat{S}^E \cup A^\Sigma$, because an unsorted axiomatization is not always given as a set of sort axioms and relativized clauses.

Hence we prefer to compare an unsorted axiomatization with its "many-sorted version", which is obtained by replacing some unary predicates by sort symbols and some axioms by an appropriate modification of the subsort order or the signature.

We illustrate this approach with the well-known "monkey-banana problem". The following proof protocol was produced by the MKRP-system for an unsorted formulation of this problem obtained from [Loveland 1978] :

```
*******************************************************************************
*                                                                             *
*    MARKGRAF KARL REFUTATION PROCEDURE, UNI KARLSRUHE, VERSION 12-OCT-83     *
*                                                                             *
*    DATE:    2-NOV-83 16:40:31                                               *
*                                                                             *
*******************************************************************************


FORMULAE GIVEN TO THE THEOREM PROVER:

AXIOMS:
     ALL X,Y ANIMAL(X) AND CLOSE.TO(X Y) IMPL CAN.REACH(X Y)
     ALL X,Y ON(X Y) AND UNDER(Y BANANA) AND TALL(Y) IMPL CLOSE.TO(X BANANA)
     ALL X,Y,Z IN.ROOM(X) AND IN.ROOM(Y) AND IN.ROOM(Z) AND
        CAN.MOVE.NEAR(X Y Z) IMPL (CLOSE.TO(Z FLOOR) OR UNDER(Y Z))
     ALL X,Y CAN.CLIMB(X Y) IMPL ON(X Y)
     ANIMAL(MONKEY)
     TALL(CHAIR)
     IN.ROOM(MONKEY)
     IN.ROOM(BANANA)
     IN.ROOM(CHAIR)
     CAN.MOVE.NEAR(MONKEY CHAIR BANANA)
     NOT CLOSE.TO(BANANA FLOOR)
     CAN.CLIMB(MONKEY CHAIR)

THEOREM:
     CAN.REACH(MONKEY BANANA)


CLAUSES:
     AXM1  :  ALL X:ANY Y:ANY NOT ANIMAL(X)  OR  NOT CLOSE.TO(X Y)
                 OR  CAN.REACH(X Y)
     AXM2  :  ALL X:ANY Y:ANY NOT ON(X Y)  OR  NOT UNDER(Y BANANA)
                 OR  NOT TALL(Y)  OR  CLOSE.TO(X BANANA)
     AXM3  :  ALL X:ANY Y:ANY Z:ANY NOT IN.ROOM(X)  OR  NOT IN.ROOM(Y)  OR
                 NOT IN.ROOM(Z) OR NOT CAN.MOVE.NEAR(X Y Z) OR CLOSE.TO(Z FLOOR)
                 OR  UNDER(Y Z)
     AXM4  :  ALL X:ANY Y:ANY NOT CAN.CLIMB(X Y)  OR  ON(X Y)
     AXM5  :  ANIMAL(MONKEY)
     AXM6  :  TALL(CHAIR)
     AXM7  :  IN.ROOM(MONKEY)
     AXM8  :  IN.ROOM(BANANA)
     AXM9  :  IN.ROOM(CHAIR)
     AXM10  :  CAN.MOVE.NEAR(MONKEY CHAIR BANANA)
     AXM11  :  NOT CLOSE.TO(BANANA FLOOR)
     AXM12  :  CAN.CLIMB(MONKEY CHAIR)
     THM13  :  NOT CAN.REACH(MONKEY BANANA)


AXM3            --> AXM3.FAC1  :  ALL X:ANY UNDER(X X)  OR  CLOSE.TO(X FLOOR)
                                     OR  NOT CAN.MOVE.NEAR(X X X)
                                     OR  NOT IN.ROOM(X)

AXM3            --> AXM3.FAC2  :  ALL X:ANY Y:ANY UNDER(X Y)  OR  CLOSE.TO(Y FLOOR)
                                     OR  NOT CAN.MOVE.NEAR(Y X Y)
                                     OR  NOT IN.ROOM(Y)  OR  NOT IN.ROOM(X)

AXM3            --> AXM3.FAC3  :  ALL X:ANY Y:ANY UNDER(X X)  OR  CLOSE.TO(X FLOOR)
                                     OR  NOT CAN.MOVE.NEAR(Y X X)  OR  NOT IN.ROOM
                                     OR  NOT IN.ROOM(X)  OR  NOT IN.ROOM(Y)

AXM3            --> AXM3.FAC4  :  ALL X:ANY Y:ANY UNDER(X Y)  OR  CLOSE.TO(Y FLOOR)
                                     OR  NOT CAN.MOVE.NEAR(X X Y)
                                     OR  NOT IN.ROOM(Y)  OR  NOT IN.ROOM(X)
```

```
AXM1  + AXM5  -->       RES1 :  ALL X:ANY CAN.REACH(MONKEY X)
                                    OR  NOT CLOSE.TO(MONKEY X)

THM13 + RES1  -->       RES2 :  NOT CLOSE.TO(MONKEY BANANA)

RES2  + AXM2  -->       RES3 :  ALL X:ANY NOT TALL(X)  OR  NOT ON(MONKEY X)
                                    OR  NOT UNDER(X BANANA)

RES3  + AXM4  -->       RES4 :  ALL X:ANY NOT UNDER(X BANANA)  OR  NOT TALL(X)
                                    OR  NOT CAN.CLIMB(MONKEY X)

RES4  + AXM12 -->       RES5 :  NOT TALL(CHAIR)  OR  NOT UNDER(CHAIR BANANA)

RES5  + AXM3  -->       RES6 :  ALL X:ANY NOT TALL(CHAIR)
                                    OR  CLOSE.TO(BANANA FLOOR)
                                    OR  NOT CAN.MOVE.NEAR(X CHAIR BANANA)
                                    OR  NOT IN.ROOM(BANANA)
                                    OR  NOT IN.ROOM(CHAIR)  OR  NOT IN.ROOM(X)

RES6  + AXM9  -->       RES7 :  ALL X:ANY NOT IN.ROOM(X)  OR  NOT IN.ROOM(BANANA)
                                    OR  NOT CAN.MOVE.NEAR(X CHAIR BANANA)
                                    OR  CLOSE.TO(BANANA FLOOR)
                                    OR  NOT TALL(CHAIR)

RES7  + AXM6  -->       RES8 :  ALL X:ANY CLOSE.TO(BANANA FLOOR)
                                    OR  NOT CAN.MOVE.NEAR(X CHAIR BANANA)
                                    OR  NOT IN.ROOM(BANANA)  OR  NOT IN.ROOM(X)

RES8  + AXM8  -->       RES9 :  ALL X:ANY NOT IN.ROOM(X)
                                    OR  NOT CAN.MOVE.NEAR(X CHAIR BANANA)
                                    OR  CLOSE.TO(BANANA FLOOR)

AXM11 + RES9  -->       RES10 :  ALL X:ANY NOT CAN.MOVE.NEAR(X CHAIR BANANA)
                                     OR  NOT IN.ROOM(X)

RES10 + AXM10 -->       RES11 :  NOT IN.ROOM(MONKEY)

RES11 + AXM7  -->       RES12 :  EMPTY


THE FOLLOWING CLAUSES WERE USED IN THE PROOF:

AXM7 AXM10 AXM8 AXM6 AXM9 AXM3 AXM12 AXM4 AXM2 AXM5 AXM1 RES1 THM13 RES2 RES3
RES4 RES5 RES6 RES7 RES8 RES9 AXM11 RES10 RES11 RES12 .


THE THEOREM IS PROVED. END OF PROOF  2-NOV-83 16:42:18.
```

Figure 12.1 The solution of the unsorted monkey-banana problem

We have chosen this problem despite the fact that it is no challenge for today's automated theorem provers. It demonstrates what advantages can be gained by many-sorted resolution even for such a simple problem with a very simple sort hierarchy. The following proof protocol was generated by the MKRP-system for a many-sorted version of the monkey-banana problem:

```
******************************************************************************
*                                                                            *
*    MARKGRAF KARL REFUTATION PROCEDURE, UNI KARLSRUHE, VERSION  12-OCT-83   *
*                                                                            *
*       DATE:    2-NOV-83 16:46:27                                           *
*                                                                            *
******************************************************************************

FORMULAE GIVEN TO THE THEOREM PROVER:

AXIOMS:
      SORT ANIMAL,TALL:IN.ROOM
      TYPE BANANA,FLOOR:IN.ROOM
      TYPE CHAIR:TALL
      TYPE MONKEY:ANIMAL
      TYPE CAN.REACH(ANIMAL IN.ROOM)
      TYPE CLOSE.TO(IN.ROOM IN.ROOM)
      TYPE ON(IN.ROOM IN.ROOM)
      TYPE UNDER(IN.ROOM IN.ROOM)
      TYPE CAN.MOVE.NEAR(ANIMAL IN.ROOM IN.ROOM)
      TYPE CAN.CLIMB(ANIMAL TALL)
      ALL X:ANIMAL ALL Y:IN.ROOM CLOSE.TO(X Y) IMPL CAN.REACH(X Y)
      ALL X:ANIMAL ALL Y:TALL ON(X Y) AND UNDER(Y BANANA) IMPL CLOSE.TO(X BANANA)
      ALL X:ANIMAL ALL Y,Z:IN.ROOM CAN.MOVE.NEAR(X Y Z)
         IMPL (CLOSE.TO(Z FLOOR) OR UNDER(Y Z)) .
      ALL X:ANIMAL ALL Y:TALL CAN.CLIMB(X Y) IMPL ON(X Y)
      CAN.MOVE.NEAR(MONKEY CHAIR BANANA)
      NOT CLOSE.TO(BANANA FLOOR)
      CAN.CLIMB(MONKEY CHAIR)

THEOREM:
      CAN.REACH(MONKEY BANANA)


CLAUSES:
      AXM1 :   ALL X:ANIMAL Y:IN.ROOM NOT CLOSE.TO(X Y) OR CAN.REACH(X Y)
      AXM2 :   ALL X:ANIMAL Y:TALL NOT ON(X Y) OR NOT UNDER(Y BANANA)
                  OR CLOSE.TO(X BANANA)
      AXM3 :   ALL X:ANIMAL Y:IN.ROOM Z:IN.ROOM NOT CAN.MOVE.NEAR(X Y Z)
                  OR CLOSE.TO(Z FLOOR) OR UNDER(Y Z)
      AXM4 :   ALL X:ANIMAL Y:TALL NOT CAN.CLIMB(X Y) OR ON(X Y)
      AXM5 :   CAN.MOVE.NEAR(MONKEY CHAIR BANANA)
      AXM6 :   NOT CLOSE.TO(BANANA FLOOR)
      AXM7 :   CAN.CLIMB(MONKEY CHAIR)
      THM8 :   NOT CAN.REACH(MONKEY BANANA)
```

```
AXM2 + AXM3 --> RES1 :  ALL X:ANIMAL Y:TALL Z:ANIMAL CLOSE.TO(X BANANA)
                            OR  NOT ON(X Y)  OR  CLOSE.TO(BANANA FLOOR)
                            OR  NOT CAN.MOVE.NEAR(Z Y BANANA)

RES1 + AXM5 --> RES2 :  ALL X:ANIMAL CLOSE.TO(BANANA FLOOR)  OR  NOT ON(X CHAIR)
                            OR  CLOSE.TO(X BANANA)

AXM1 + RES2 --> RES3 :  ALL X:ANIMAL CAN.REACH(X BANANA)  OR  NOT ON(X CHAIR)
                            OR  CLOSE.TO(BANANA FLOOR)

AXM6 + RES3 --> RES4 :  ALL X:ANIMAL NOT ON(X CHAIR)  OR  CAN.REACH(X BANANA)

THM8 + RES4 --> RES5 :  NOT ON(MONKEY CHAIR)

RES5 + AXM4 --> RES6 :  NOT CAN.CLIMB(MONKEY CHAIR)

RES6 + AXM7 --> RES7 :  EMPTY




THE FOLLOWING CLAUSES WERE USED IN THE PROOF:

AXM7 AXM4 AXM5 AXM3 AXM2 RES1 RES2 AXM1 RES3 AXM6 RES4 THM8 RES5 RES6 RES7 .


THE THEOREM IS PROVED. END OF PROOF  2-NOV-83 16:47:22.
```

Figure 12.2 The solution of the many-sorted monkey-banana problem

The following diagram compares the statistics of both runs. The value for **initial links** corresponds to the size of the initial search space and the value for **links generated** is a measure for the whole search space generated during a proof run. (Note that a connection graph link represents a potential deduction of a clause.) The value for **steps executed** corresponds to the costs of the actual search and **level of proof** is the search depth. The values for **R-links**, **P-links** and **F-links** are the numbers of potential resolvents, paramodulants and factors, i.e. clauses which could be deduced during the proof run. The first column gives the value for the many-sorted solution, and the second column that for the unsorted one. The ratios between the corresponding values are given in per cent in column 3 and graphically in column 4 (where the black portion in the boxes decreases with an increasing advantage of the many-sorted solution).

	ΣRP	RP	$\Sigma RP/_{RP}$ *100
CPU seconds	3.3	11.4	29
steps executed	7	16	44
links generated	22	99	22
R-links	22	95	23
P-links	0	0	100
F-links	0	4	0
initial links	8	23	35
R-links	8	19	42
P-links	0	0	100
F-links	0	4	0
clauses generated	15	29	52
initial	8	13	62
deduced	7	16	44
resolvents	7	12	58
paramodulants	0	0	100
factors	0	4	0
literals generated	28	75	37
initial	14	24	58
deduced	14	51	28
level of proof	7	12	58
clauses in proof	15	25	60

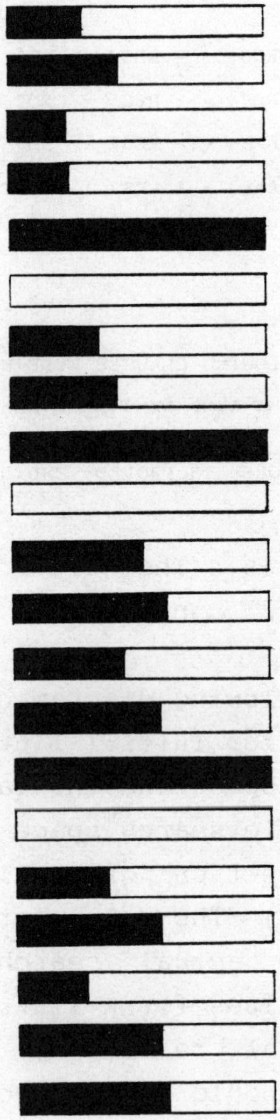

Figure 12.3 The proof statistics for the monkey-banana problem

The statistical values in Figure 12.3 reveal the typical advantages of the ΣRP-calculus in automated theorem proving, viz. a drastic reduction of the search space, which entails a reduction of the search depth and of the costs for a search. Note that for this example the reduction of the search space is caused only by the reduction of the initial clause set. Generally, i.e. if the sort hierarchy used is more complex, an additional reduction is obtained by the sortal constraints imposed on the unification procedure.

We shall illustrate this effect with another example, which was a challenge for the MKRP-system for several years. In 1978, L. Schubert from the University of Alberta set up the following reasoning problem (which became known as "Schubert's Steamroller"):

Wolves, foxes, birds, caterpillars, and snails are animals, and there are some of each of them. Also there are some grains and grains are plants. Every animal either likes to eat all plants or all animals much smaller than itself that like to eat some plants. Caterpillars and snails are much smaller than birds, which are much smaller than foxes, which in turn are much smaller than wolves. Wolves do not like to eat foxes or grains, while birds like to eat caterpillars but not snails. Caterpillars and snails like to eat some plants. Therefore there is an animal that likes to eat a grain-eating animal.

Using an unsorted axiomatization of the problem, the MKRP-system was unable to find a solution (temporarily). But with a many-sorted axiomatization a solution was computed without difficulty [Walther 1984b].

```
*************************************************************************************
*                                                                                   *
*   MARKGRAF KARL REFUTATION PROCEDURE, UNI KARLSRUHE, VERSION 31-OCT-83            *
*                                                                                   *
*   DATE:  20-DEC-83 12:52:23                                                       *
*                                                                                   *
*************************************************************************************


FORMULAE GIVEN TO THE THEOREM PROVER:

AXIOMS:
     TYPE W:WOLF
     TYPE F:FOX
     TYPE B:BIRD
     TYPE C:CATERPILLAR
     TYPE S:SNAIL
     TYPE G:GRAIN
     SORT WOLF,FOX,BIRD,CATERPILLAR,SNAIL:ANIMAL
     SORT GRAIN:PLANT
     ALL W:ANIMAL (ALL X:PLANT EATS(W X)) OR (ALL Y:ANIMAL SMALLER(Y W) AND
        (EX Z:PLANT EATS(Y Z)) IMPL EATS(W Y))
     ALL X:CATERPILLAR ALL Y:BIRD SMALLER(X Y)
     ALL X:SNAIL ALL Y:BIRD SMALLER(X Y)
     ALL X:BIRD ALL Y:FOX SMALLER(X Y)
     ALL X:FOX ALL Y:WOLF SMALLER(X Y)
     ALL X:WOLF ALL Y:FOX NOT EATS(X Y)
     ALL X:WOLF ALL Y:GRAIN NOT EATS(X Y)
     ALL X:BIRD ALL Y:CATERPILLAR EATS(X Y)
     ALL X:BIRD ALL Y:SNAIL NOT EATS(X Y)
     ALL X:CATERPILLAR EX Y:PLANT EATS(X Y)
     ALL X:SNAIL EX Y:PLANT EATS(X Y)

THEOREM:
     EX X,Y:ANIMAL (ALL Z:GRAIN EATS(X Y) AND EATS(Y Z))


SKOLEM SYMBOLS:
     TYPE F_1(CATERPILLAR):PLANT
     TYPE F_2(SNAIL):PLANT
     TYPE F_3(ANIMAL ANIMAL):GRAIN

CLAUSES:
     AXM1  :  ALL W:ANIMAL X:PLANT Y:ANIMAL Z:PLANT EATS(W X)  OR
                 NOT SMALLER(Y W)  OR  NOT EATS(Y Z)  OR  EATS(W Y)
     AXM2  :  ALL X:CATERPILLAR Y:BIRD SMALLER(X Y)
     AXM3  :  ALL X:SNAIL Y:BIRD SMALLER(X Y)
     AXM4  :  ALL X:BIRD Y:FOX SMALLER(X Y)
     AXM5  :  ALL X:FOX Y:WOLF SMALLER(X Y)
     AXM6  :  ALL X:WOLF Y:FOX NOT EATS(X Y)
     AXM7  :  ALL X:WOLF Y:GRAIN NOT EATS(X Y)
     AXM8  :  ALL X:BIRD Y:CATERPILLAR EATS(X Y)
     AXM9  :  ALL X:BIRD Y:SNAIL NOT EATS(X Y)
     AXM10  :  ALL X:CATERPILLAR EATS(X F_1(X))
     AXM11  :  ALL X:SNAIL EATS(X F_2(X))
     THM12  :  ALL X:ANIMAL Y:ANIMAL NOT EATS(X Y)  OR  NOT EATS(Y F_3(X Y))


THM12 + AXM8  -->       RES1 :  ALL X:BIRD Y:CATERPILLAR NOT EATS(Y F_3(X Y))

THM12 + AXM1  -->       RES2 :  ALL X:ANIMAL Y:PLANT Z:ANIMAL EATS(X Y)
                                    OR NOT SMALLER(Z X)  OR  NOT EATS(Z F_3(X Z))

RES2  + AXM5  -->       RES3 :  ALL X:WOLF Y:PLANT Z:FOX EATS(X Y)  OR
                                    NOT EATS(Z F_3(X Z))

AXM7  + RES3  -->       RES4 :  ALL X:FOX Y:WOLF NOT EATS(X F_3(Y X))
```

```
RES2  + AXM4  -->      RES5 :  ALL X:FOX Y:PLANT Z:BIRD EATS(X Y)  OR
                                  NOT EATS(Z F_3(X Z))

RES4  + RES5  -->      RES6 :  ALL X:BIRD Y:FOX NOT EATS(X F_3(Y X))

AXM9  + AXM1  -->      RES7 :  ALL X:BIRD Y:PLANT Z:SNAIL W:PLANT EATS(X Y)
                                  OR  NOT EATS(Z W)  OR  NOT SMALLER(Z X)

RES7  + RES6  -->      RES8 :  ALL X:SNAIL Y:BIRD Z:PLANT NOT EATS(X Z)
                                  OR  NOT SMALLER(X Y)

RES8  + AXM3  -->      RES9 :  ALL X:SNAIL Y:PLANT NOT EATS(X Y)

AXM11 + RES9  -->      RES1Ø :  EMPTY


THE FOLLOWING CLAUSES WERE USED IN THE PROOF:

AXM11 AXM3 AXM1 AXM4 THM12 RES2 RES5 AXM5 RES3 AXM7 RES4 RES6 RES7 RES8 RES9
AXM9 RES1Ø .


THE THEOREM IS PROVED. END OF PROOF  2Ø-DEC-83 12:53:Ø8.
```

Figure 12.4 A many-sorted solution of Schubert's Steamroller

But why did the system find a solution for the many-sorted formulation, when it did not find one for the unsorted type? The reason becomes apparent if we compare the statistics of the many-sorted solution with the values for Schubert's hand-computed deduction of the empty clause [Schubert 1978, Walther 1984a] (where the value for initial links is obtained from the unsuccessful run of the MKRP-system).

	ΣRP	RP	ΣRP/RP *100
CPU seconds	7.1	-	-
steps executed	10	-	-
links generated	46	-	-
R-links	46	-	-
P-links	0	-	-
F-links	0	-	-
initial links	12	102	12
R-links	12	94	13
P-links	0	0	100
F-links	0	8	0
clauses generated	22	60	37
initial	12	27	44
deduced	10	33	30
resolvents	10	33	30
paramodulants	0	0	100
factors	0	0	100
literals generated	32	221	15
initial	16	65	25
deduced	16	156	10
level of proof	8	20	40
clauses in proof	17	54	32

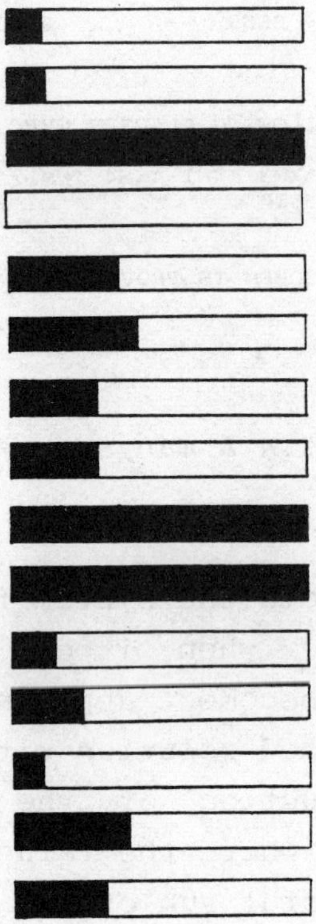

Figure 12.5 The proof statistics for Schubert's Steamroller

In the unsorted case, the size of the initial search space (we can compute 94 resolvents and 8 factors already in the first generation) and the search depth necessary to compute the empty clause (which is 20 in the hand-made solution) leads to such a rapidly growing search space, that the time and/or space boundaries of an automated theorem prover are exceeded before the empty clause is found, unless special reasoning techniques are used. This holds true even if we use some refinements, like, for instance, set-of-support, cf. [Chang and Lee 1973, Loveland 1978], which reduces the initial search space to 28 potential resolvents and 2 potential factors. For this example, the reduction of the search space is also caused by the constraints imposed on many-sorted unification (cf. the discussion of Example 2.1 in Chapter 2).

After several significant improvements, in particular of the terminator module [Antoniou and Ohlbach 1983], the MKRP-system was also able to find a refutation of the unsorted problem. Using 261.7 seconds running time, the system computed a proof of 76 clauses after 87 clauses had been generated. Obviously, Schubert's Steamroller causes combinatorical difficulties for a theorem proving system, and this may be the reason why this problem became a criterion to verify the performance of several automated theorem provers [Stickel 1986].

It seems that the advantage of the ΣRP-calculus in automated theorem proving increases (or decreases) with an increasing (or decreasing) complexity of the sort hierarchies under consideration. But this holds true only if we restrict ourselves to many-sorted resolution. Consider, for instance, a sort hierarchy $(\mathcal{S},<_{\mathcal{S}})$ where $<_{\mathcal{S}}$ is empty. Then each most general unifier of some complementary pair of literals is a well-sorted most general unifier (because all literals are well-formed and no subsorts exist). Hence each resolvent is a many-sorted one, because there are no constraints imposed on unification for this sort hierarchy. Since with $<_{\mathcal{S}}=\emptyset$ there would be no sort axioms for the subsort order in an unsorted axiomatization, we can take advantage only from the sort axioms belonging to the signature.

But the situation is completely different for many-sorted paramodulation, because we also have to unify terms in this case. Let us assume that some clause of a given clause set S contains an equality literal x≡q. Then each term t in each clause of S can be replaced by q in a paramodulation step, because {x,t} is always unifiable. But for the many-sorted case, the unification of {x,t} depends on the sorts of x and t. Using a sort hierarchy with $<_{\mathcal{S}}=\emptyset$, for instance, the constraints imposed on many-sorted unification demand that x and t have equal sorts. Hence we obtain a strong restriction for the computation of many-sorted paramodulants in this case, i.e. a strong reduction of the search space.

We shall illustrate this effect with a simple example. Given some sets A and B and some mappings $f:A\times A\rightarrow A$, $g:B\times B\rightarrow B$ and $h:A\rightarrow B$ such that h(f(xy))=g(h(x)h(y)) for all x,y∈A, we want to prove that g(zz)=z for all z∈h(A), if f(xx)=x for all x∈A. The following proof protocol was generated by the MKRP-system for a many-sorted formulation of the problem:

```
*****************************************************************************
*                                                                           *
*    MARKGRAF KARL REFUTATION PROCEDURE, UNI KARLSRUHE, VERSION  12-OCT-83  *
*                                                                           *
*    DATE:  14-NOV-83 15:47:11                                              *
*                                                                           *
*****************************************************************************


FORMULAE GIVEN TO THE THEOREM PROVER:

AXIOMS:
     TYPE F(A A):A
     TYPE G(B B):B
     TYPE H(A):B
     ALL X,Y:A H(F(X Y)) = G(H(X) H(Y))

THEOREM:
     (ALL X:A F(X X) = X) IMPL (ALL Y:A G(H(Y) H(Y)) = H(Y))


SKOLEM SYMBOLS:
     TYPE C_1:A

CLAUSES:
     AXM1  :  ALL X:A Y:A H(F(X Y))=G(H(X) H(Y))
     THM2  :  ALL X:A F(X X)=X
     THM3  :  NOT G(H(C_1) H(C_1))=H(C_1)


THM3  + AXM1  -->      PAR1  :  NOT H(F(C_1 C_1))=H(C_1)

PAR1  + THM2  -->      PAR2  :  NOT H(C_1)=H(C_1)

                       AXMØ  :  ALL X:ANY X=X

PAR2  + AXMØ  -->      RES1  :  EMPTY


THE FOLLOWING CLAUSES WERE USED IN THE PROOF:

AXMØ THM2 AXM1 THM3 PAR1 PAR2 RES1 .

THE THEOREM IS PROVED. END OF PROOF  14-NOV-83 15:48:18.
```

Figure 12.6 The solution of the many-sorted idempotent-problem

If we ignore the signature for the function symbols f,g and h, we obtain the same refutation in the RP-calculus:

```
*****************************************************************************************
*                                                                                       *
*     MARKGRAF KARL REFUTATION PROCEDURE, UNI KARLSRUHE, VERSION 31-OCT-83              *
*                                                                                       *
*     DATE:  14-NOV-83 15:50:02                                                         *
*                                                                                       *
*****************************************************************************************


FORMULAE GIVEN TO THE THEOREM PROVER:

AXIOMS:
     ALL X,Y H(F(X Y)) = G(H(X) H(Y))

THEOREM:
     (ALL X F(X X) = X) IMPL (ALL Y G(H(Y) H(Y)) = H(Y))


CLAUSES:
     AXM1  :  ALL X:ANY Y:ANY H(F(X Y))=G(H(X) H(Y))
     THM2  :  ALL X:ANY F(X X)=X
     THM3  :  NOT G(H(C_1) H(C_1))=H(C_1)


THM3  + AXM1  -->       PAR1  :  NOT H(F(C_1 C_1))=H(C_1)

PAR1  + THM2  -->       PAR2  :  NOT H(C_1)=H(C_1)

                        AXM0  :  ALL X:ANY X=X

PAR2  + AXM0  -->       RES1  :  EMPTY


THE FOLLOWING CLAUSES WERE USED IN THE PROOF:

AXM0 THM2 AXM1 THM3 PAR1 PAR2 RES1 .


THE THEOREM IS PROVED. END OF PROOF  14-NOV-83 15:53:15.
```

Figure 12.7 The solution of the idempotent-problem without sorts

The following diagram compares the statistics of both solutions:

	ΣRP	RP	$\Sigma RP/_{RP}$ *100
CPU seconds	1	1.4	74
steps executed	3	3	100
links generated	21	36	58
R-links	1	1	100
P-links	20	35	57
F-links	0	0	100
initial links	14	26	54
R-links	0	0	100
P-links	14	26	54
F-links	0	0	100
clauses generated	6	6	100
initial	3	3	100
deduced	3	3	100
resolvents	1	1	100
paramodulants	2	2	100
factors	0	0	100
literals generated	6	6	100
initial	3	3	100
deduced	3	3	100
level of proof	3	3	100
clauses in proof	7	7	100

Figure 12.8 The proof statistics for the many-sorted idempotent-problem and a formulation without sorts

As expected, there is no increase in potential resolution steps if we ignore the sorts as in Figure 12.7. But there is a strong increase in potential paramodulations. The reason is that in the formulation without sorts, each term of form g(...) or h(...) is unifiable with the right hand side of f(xx)≡x, whereas in the many-sorted case those unifications are impossible, because the sorts of the terms involved do not agree.

But even worse, we did not solve the given problem, because we have identified the sets A and B in the formulation without sorts. Hence we have to relativize the formulas of the idempotent-problem, to obtain an equivalent unsorted formulation:

```
*****************************************************************************
*                                                                           *
*    MARKGRAF KARL REFUTATION PROCEDURE, UNI KARLSRUHE, VERSION 31-OCT-83   *
*                                                                           *
*    DATE:  14-NOV-83 15:54:54                                              *
*                                                                           *
*****************************************************************************


FORMULAE GIVEN TO THE THEOREM PROVER:

AXIOMS:
     ALL X,Y A(X) AND A(Y) IMPL H(F(X Y)) = G(H(X) H(Y))

THEOREM:
     (ALL X A(X) IMPL F(X X) = X) IMPL (ALL Y A(Y) IMPL G(H(Y) H(Y)) = H(Y))

CLAUSES:
     AXM1 :   ALL X:ANY Y:ANY NOT A(X)  OR  NOT A(Y)  OR  H(F(X Y))=G(H(X) H(Y))
     THM2 :   ALL X:ANY NOT A(X)  OR  F(X X)=X
     THM3 :   A(C_1)
     THM4 :   NOT G(H(C_1) H(C_1))=H(C_1)


THM2  + THM3  -->      RES1 :  F(C_1 C_1)=C_1

THM4  + RES1  -->      PAR1 :  NOT G(H(C_1) H(C_1))=H(F(C_1 C_1))

AXM1  + THM3  -->      RES2 :  H(F(C_1 C_1))=G(H(C_1) H(C_1))  OR  NOT A(C_1)

RES2  + THM3  -->      RES3 :  H(F(C_1 C_1))=G(H(C_1) H(C_1))

PAR1  + RES3  -->      RES4 :  EMPTY


THE FOLLOWING CLAUSES WERE USED IN THE PROOF:

THM3 AXM1 RES2 RES3 THM2 RES1 THM4 PAR1 RES4 .


THE THEOREM IS PROVED. END OF PROOF  14-NOV-83 15:56:01.
```

Figure 12.9 The solution of the unsorted idempotent-problem

With an equivalent unsorted formulation of the problem, the search space increases significantly, since by the presence of sort literals many additional resolvents and paramodulants can be deduced:

	ΣRP	RP	$\Sigma RP/_{RP}$ *100
CPU seconds	1	3.7	28
steps executed	3	5	60
links generated	21	119	18
R-links	1	7	14
P-links	20	112	18
F-links	0	1	0
initial links	14	37	38
R-links	0	3	0
P-links	14	33	42
F-links	0	1	0
clauses generated	6	9	67
initial	3	4	75
deduced	3	5	60
resolvents	1	4	25
paramodulants	2	1	200
factors	0	0	100
literals generated	6	12	50
initial	3	7	43
deduced	3	5	60
level of proof	3	3	100
clauses in proof	7	9	78

Figure 12.10 The statistics for the many-sorted and the unsorted idempotent-problem

13 Related Work and Concluding Remarks

In his thesis, J. Herbrand presented a many-sorted version of his calculus and proved the prooftheoretic part of the Sort-Theorem [Herbrand 1930]. However Herbrand's proof is inadequate, because he did not consider that certain deductions in his one-sorted calculus cannot be translated to deductions in the many-sorted calculus. This was pointed out by A. Schmidt [Schmidt 1938], who proposed a many-sorted version of a Hilbert-type calculus without subsorts and proved the prooftheoretic part of the Sort-Theorem for this calculus [Schmidt 1938, Schmidt 1951].

H. Wang defined a many-sorted version of a Hilbert-type calculus without function symbols and subsorts [Wang 1952]. He proved the soundness and completeness of his calculus and the modeltheoretic part of the Sort-Theorem. Wang also gave an alternative proof of the prooftheoretic part of the Sort-Theorem by an application of the Herbrand-Theorem.

P. C. Gilmore pointed out that this proof was inadequate. He extended the many-sorted calculus of Wang by the introduction of subsorts and presented an improved version of the prooftheoretic part of the Sort-Theorem for this extended calculus [Gilmore 1958].

T. Hailperin proposed a calculus which can be viewed as a generalization of Wang's many-sorted calculus [Hailperin 1957]. In this calculus sortrelationships can be expressed using arbitrary first-order formulas instead of using atomar formulas (corresponding to sort literals). Hailperin proved a theorem which corresponds to the prooftheoretic part of the Sort-Theorem.

A. Oberschelp proposed several many-sorted versions of a calculus of Montague and Henkin [Montague and Henkin 1956]. In these calculi function symbols and subsorts are admitted. Oberschelp proved the soundness and completeness of his calculi and also gave the proofs for the modeltheoretic parts of the Sort-Theorems [Oberschelp 1962].

A. V. Idelson discussed forms of many-sorted calculi of constructive mathematical logic [Idelson 1964], which are based on the calculus of natural deduction [Gentzen 1934].

With the emerging field of automated theorem proving a first-order calculus became a practical tool to find mathematical proofs. Thereby the works cited above received practical significance. The advantages of many-sorted reasoning were also recognized within this field, e.g. [Hayes 1971, Henschen 1972], and several theorem proving programs have been based on some kind of a many-sorted calculus, e.g. [Weyhrauch 1977, Champeaux 1978] (unfortunately without a sound theoretical foundation).

Based on an early version proposed in [Walther 1983], Schmidt-Schauss extended the ΣRP-calculus with polymorphic function symbols (where the sort of a term $f(t_1 \ldots t_n)$ can vary depending on the sorts of its arguments $t_1, \ldots, t_n$). He investigated many-sorted unification with polymorphic functions and proved the soundness and the completeness and also the modeltheoretic part of the Sort-Theorem for his extension [Schmidt-Schauss 1985a]. A further extension was obtained by incorporation of unification under equational theories into the many-sorted system [Schmidt-Schauss 1986]. Also an algorithm has been developed, which translates an unsorted clause set into a many-sorted one, where certain unary predicates from the unsorted formulation are used as sort symbols in the many-sorted version [Schmidt-Schauss 1985b].

Polymorphism is also a concern of A. G. Cohn in his proposal of a many-sorted logic, which is proved to be sound and complete [Cohn 1983]. But more important, Cohn's system differs from other many-sorted logics in the implicit assignment of sorts to variables (depending on the argument position the variable occupies). Consequently, the sort of a term need not to be a subsort of or equal to the domainsort of the argument position it occupies. Also sort literals, which are usually supplanted by the presence of sorts, may be used in his many-sorted logic. All these features entail an increase of expressiveness. Hence statements which would normally require a well-sorted formula, can be represented implicitly using the sortal apparatus of Cohn's calculus. The inference machinery of the system (which, of course, is much more complicated compared to a pure resolution based system) can take advantage of these additional information, e.g. by "evaluation" of literals, and deduction can become very effective [Cohn 1985].

Many-sorted reasoning has also found attention in the field of term rewriting systems and equality reasoning using the Knuth-Bendix completion algorithm [Knuth and Bendix 1970, Huet and Oppen 1980]. In this approach conditional equations are difficult to treat and therefore the use of sorts (and subsorts) can be advantageous because conditions can sometimes be expressed by sortal constraints. A further advantage is pointed out in [Cunningham and Dick 1985]: sorts can be used to deal economically with partial functions without leaving the scope of equational reasoning or introducing excessive formal clutter by error elements. Cunningham and Dick propose a lattice of types, which gives a natural framework for specifying data types. A unification algorithm is presented and it is shown that complete and minimal sets of Σ-unifiers can have more than one member, if $(\mathcal{S}, <_{\mathcal{S}})$ is a lattice and polymorphic functions are admitted. An algorithm for Σ-matching is given which returns at most one substitution. Goguen et al. present rewrite rule-based operational semantics for many-sorted algebra with subsorts and polymorphism and discuss pattern matching as well as term rewriting in this context [Goguen et al. 1985].

Frisch gives a short survey of advantages which sortal inferences by many-sorted unification may yield in logic programming, in parsing with Horne-clause grammars and in knowledge retrieval [Frisch 1985a, Frisch 1985b]. Huber and Varsek discuss the advantages and consequences of many-sorted resolution in logic programming [Huber and Varsek 1986]. Ait-Kaci and Nasr rediscover many-sorted unification and related concepts and discuss the advantages of many-sorted logic in a PROLOG environment [Ait-Kaci and Nasr 1986].

° ° °

For the RP-calculus many refinements like set-of-support, linear resolution, hyperresolution etc. have been proposed and their behaviour regarding completeness was investigated, cf. [Chang and Lee 1973, Loveland 1978]. Instead of reinvestigating these results for the ΣRP-calculus, is seems advantageous to have a direct proof of the prooftheoretic part of the Sort-Theorem 10.7 (ii), i.e.

$$\hat{S}^E \cup A^\Sigma \vdash \square \text{ iff } S^E \vdash_\Sigma \square \quad ,$$

for each $\mathcal{S}$-sorted signature Σ and each Σ-clause set S. This direct proof would provide a constructive way to translate a refutation in the RP-calculus uniformly to a refutation in the ΣRP-calculus (whenever this is possible). We conjecture that such a construction would show (in most cases), whether completeness results for refinements in the RP-calculus also hold for the ΣRP-calculus.

References

AIT-KACI, H. and NASR, R. LOGIN: A Logic Programming Language with Built-in Inheritance. J. Logic Programming, vol 3, no 3, 1986

ANTONIOU, G. and OHLBACH, H.J. Terminator. Proceedings of the 8th International Joint Conference on Artificial Intelligence (IJCAI-83), Karlsruhe, 1983

BIRKHOFF, G. Lattice Theory. American Mathematical Society Colloquium Publications, vol xxv, 1967.

BLÄSIUS, K., EISINGER, N., SIEKMANN, J., SMOLKA, G., HEROLD, A. and WALTHER, C. The Markgraf Karl Refutation Procedure (Fall 1981). Proceedings of the 7th International Joint Conference on Artificial Intelligence (IJCAI-81), Vancouver, 1981

BRAND, D. Proving Theorems with the Modification Method. SIAM J. Computing 4, 1975

CHAMPEAUX, D. de A Theorem Prover Dating a Semantic Network. Proceedings of the AISB/GI Conference, Hamburg, 1978

CHANG, C.-L. and LEE, R.C.-T. Symbolic Logic and Mechanical Theorem Proving. Academic Press, 1973

COHN, A.G. Improving the Expressiveness of Many-Sorted Logic. Proceedings of the 3rd National Conference on Artificial Intelligence (AAAI-83), Washington, 1983

COHN, A.G. On the Solution of Schubert's Steamroller in Many Sorted Logic. Proceedings of the 9th International Conference on Artificial Intelligence (IJCAI-85), Washington, 1985

COHN, P.M. Universal Algebra. Reidel, Dordrecht, 1981

CUNNINGHAM, R.J. and DICK, A.J.J. Rewrite systems on a lattice of types. Acta Informatica, vol 22, no 2, 1985

EDER, E. Properties of Substitutions and Unifications. J. Symbolic Computation, vol 1, no 1, 1985

EISINGER, N. and OHLBACH, H.J. The Markgraf Karl Refutation Procedure (MKRP). Proceedings of the 8th International Conference on Automated Deduction (CADE-86), Lecture Notes in Computer Science 230, Oxford, 1986

EISINGER, N., SIEKMANN, J. and WRIGHTSON, G. Paramodulated Connectiongraphs. Proceedings of the AISB/GI Conference on Artificial Intelligence, Hamburg, 1978

FAGES, F. and HUET, D. Complete Sets of Unifiers and Matchers in Equational Theories. Proceedings of the 8th CAAP, Lecture Notes in Computer Science 159, L'Aquila, 1983

FRISCH, A.M. An Investigation into Inference with Restricted Quantification and a Taxonomic Representation. SIGART Newsletter 91, 1985 (a)

FRISCH, A.M. Parsing with Restricted Quantification. Proceedings of the AISB Conference, 1985 (b)

GENTZEN, G. Untersuchungen über das logische Schließen. Mathematische Zeitschrift 39, 1934

GILMORE, P.C. An Addition to "Logic of Many-Sorted Theories". Composito Mathematica 13, 1958

GOGUEN, J.A., JOUANNAUD, J.-P., and MESEGUER, J. Operational Semantics for Order-Sorted Algebra. Proceedings of the 12th ICALP, Lecture Notes in Computer Science 194, Nafplion, 1985

GOGUEN, J.A., THATCHER, J.W. and WAGNER, E.G. An Initial Algebra Approach to the Specification, Correctness and Implementation of Abstract Data Types. IBM Research Report RC 6487, 1977, also in: Current Trends in Programming Methodology, vol 4, Data Structuring. R.T. Yeh, Ed., Prentice Hall,1978

HAILPERIN, T. A Theory of Restricted Quantification I. J. Symbolic Logic 22, 1957

HAYES, P. A Logic of Actions. Machine Intelligence 6, 1971

HENSCHEN, L.J. N-Sorted Logic for Automatic Theorem Proving in Higher Order Logic. Proceedings of the ACM Conference, Boston, 1972

HERBRAND, J. Recherches sur la théorie de la démonstration (Thèse Paris), Warsaw, 1930, Chapter 3, also in: Jacques Herbrand - Locical Writings. W.D. Goldfarb, Ed., Reidel, Dordrecht, 1971

HEROLD, A. Some basic Notions of First-Order Unification Theory. Interner Bericht 15/83, Universität Karlsruhe, Institut für Informatik I, 1983

HUBER, M. and VARSEK, I. Extended Prolog for Order-Sorted Resolution (EPOS). Interner Bericht, Institut für Informatik I, Universität Karlsruhe, 1986 (forthcoming)

HUET, G. and OPPEN, D.C. Equations and Rewrite Rules: A Survey. Technical Report CSL-111, SRI International, 1980, also in: Formal Languages Theory: Perspectives and Open Problems, R. Book, Ed., Academic Press, 1980

IDELSON, A.V. Calculi of Constructive Logic with Subordinate Variables. American Mathematical Society Translations (2) 99, 1972 - translation of Trudy Mat. Inst. Steklov 72, 1964

IRANI, K.B. and SHIN, D.G. A Many-Sorted Resolution based on an Extension of a First-Order Language. Proceedings of the 9th International Joint Conference on Artificial Intelligence (IJCAI-85), Los Angeles, 1985

KNUTH, D. and BENDIX, P. Simple Word Problems in Universal Algebras. In: Computational Problems in Abstract Algebra. J. Leech, Ed., 1970

KOWALSKI, R. A Proof Procedure Using Connection Graphs. J. ACM, vol 22, no. 4, 1975

LOVELAND, D.W. Automated Theorem Proving: A Logical Basis. North-Holland Publishing Company, 1978

MACNEILLE, H.M. Partially Ordered Sets. Transactions American Mathematical Society 42, 1937

MENDELSON, E. Introduction to Mathematical Logic. D. Van Nostrand Company, New York, 1964

MONTAGUE, R. and HENKIN, L. On the Definition of Formal Deduction. J. Symbolic Logic 21, 1956

OBERSCHELP, A. Untersuchungen zur mehrsortigen Quantorenlogik. Mathematische Annalen 145, 1962

PLOTKIN, G. Building in Equational Theories. Machine Intelligence 7, 1972

ROBINSON, J.A. A Machine-Oriented Logic Based on the Resolution Principle. J. ACM, vol 12, no 1, 1965, also in: Automation of Reasoning - Classical Papers on Computational Logic, vol 1. J. Siekmann and G. Wrightson, Eds., Springer, 1983.

ROSEN, B.K. Tree-Manipulating Systems and Church-Rosser Theorems. J. ACM, vol 20, no 1, 1973

SCHMIDT, A. Über deduktive Theorien mit mehreren Sorten von Grunddingen. Mathematische Annalen 115, 1938

SCHMIDT, A. Die Zulässigkeit der Behandlung mehrsortiger Theorien mittels der üblichen einsortigen Prädikatenlogik. Mathematische Annalen 123, 1951

SCHMIDT-SCHAUSS, M. A Many-Sorted Calculus with Polymorphic Functions based on Resolution and Paramodulation. Proceedings of the 9th International Joint Conference on Artificial Intelligence (IJCAI-85), Los Angeles, 1985 (a)

SCHMIDT-SCHAUSS, M. Mechanical Generation of Sorts in Clause Sets. MEMO SEKI-85-VI-KL, Fachbereich Informatik, Universität Kaiserslautern, 1985 (b)

SCHMIDT-SCHAUSS, M. Unification in Many-Sorted Equational Theories. Proceedings of the 8th International Conference on Automated Deduction (CADE-86), Lecture Notes in Computer Science 230, Oxford, 1986

SCHUBERT, L. Personal communication, 1978

SIEKMANN, J. Universal Unification. Proceedings of the 7th International Conference on Automated Deduction (CADE-84), Lecture Notes in Computer Science 170, Napa, 1984

SIEKMANN, J. and WRIGHTSON, G. Paramodulated Connectiongraphs. Acta Informatica 13, 1980

STICKEL, M. E. Schubert's Steamroller Problem: Formulations and Solutions. J. Automated Reasoning, vol 2, no 1, 1986

WALTHER, C. Elimination of Redundant Links in Extended Connection Graphs. Proceedings of the 5th German Workshop on Artificial Intelligence (GWAI-81), Springer Fachberichte Informatik 47, Bad Honnef, 1981

WALTHER, C. The Markgraf Karl Refutation Procedure: PLL - A First-Order Language for an Automated Theorem Prover. Interner Bericht 35/82, Institut für Informatik I, Universität Karlsruhe, 1982

WALTHER, C. A Many-Sorted Calculus Based on Resolution and Paramodulation. Proceedings of the 8th International Joint Conference on Artificial Intelligence (IJCAI-83), Karlsruhe, 1983

WALTHER, C. Schubert's Steamroller - A Case Study in Many-Sorted Resolution. Interner Bericht 5/84, Institut für Informatik I, Universität Karlsruhe, 1984 (a)

WALTHER, C. A Mechanical Solution of Schubert's Steamroller by Many-Sorted Resolution. Proceedings of the 4th National Conference on Artificial Intelligence (AAAI-84), Austin, 1984 (b), revised version in: Artificial Intelligence vol 26, no 2, 1985

WALTHER, C. Unification in Many-Sorted Theories. Proceedings of the 6th European Conference on Artificial Intelligence (ECAI-84), Pisa, 1984 (c), also in: Advances in Artificial Intelligence. T. O'Shea, Ed., North-Holland, 1985

WALTHER, C. A Classification of Many-Sorted Unification Problems. Proceedings of the 8th International Conference on Automated Deduction (CADE-86), Lecture Notes in Computer Science 230, Oxford, 1986

WANG, H. Logic of Many-Sorted Theories. J. Symbolic Logic 17, 1952

WEYHRAUCH, R.W. FOL: A Proof Checker for First-Order Logic. MEMO AIM-235.1, Stanford Artificial Intelligence Laboratory, Stanford University, 1977

WOS, L. and ROBINSON, G. Maximal Models and Refutation Completeness: Semidecision Procedures in Automatic Theorem Proving. In: Wordproblems. W.W. Boone, F.B. Cannonito, and R.C. Lyndon, Eds., North-Holland, 1973, also in: Automation of Reasoning - Classical Papers on Computational Logic, vol 2, J. Siekmann and G. Wrightson, Eds., Springer, 1983